Mysticism, which looms so large today, was excluded from the history of science and medicine on the grounds that it was irrelevant to the mainstream of scientific thought. Positivist historians of science directed their attention to that segment of the past that could be related to present concerns and contemporary methods. In recent years, however, doubt has been cast on the overall achievement of modern science and, consequently, the clarity of the distinction between wrong and right steps, between the false starts and the successful arrivals in the history of science has been blurred. This has made it easier for historians to take a further step in reappraising the historiography of science and to raise the crucial question: are "mystical" traditions and "modern" elements in the Scientific Revolution antagonistic or, paradoxically, complementary extremes?

The challenging essays in this volume are reports of work in progress. While it is too early to discern the complete picture of the method of seventeenth-century science, they make clear that hermeticism and alchemy contributed positively to the development of the experimental method by stressing the importance of observation, freeing science from the weight of inherited authority, recognizing the value and dignity of the crafts, and, perhaps most significantly, by emphasizing the utilitarian goal of scientific knowledge.

This new method was not elaborated in one day, nor in one century, related as it was to the entire *Weltanschauung* of Western Civilization, but the gradual and often chaotic transformation of what constitutes scientific knowledge should not blind us to the realization that we owe our new outlook on the world to this golden age.

Luisa Righini Bonelli is the Direc- ne Institute and the Museum of the of Science in Florence and the of several books and over two d articles on the history of scien- struments, the School of Galileo, Accademia del Cimento. She is e of the astronomer, Guglielmo , who also contributed an article volume.

R. Shea, the author of *Galileo's tual Revolution* and several arti- he history and philosophy of science, is Associate Professor of the History and Philosophy of Science at McGill University. In 1973-1974, Professor Shea was Visiting Professor at the Institute of the History of Science and a Fellow of the Harvard University Renaissance Center in Florence.

Reason, Experiment, and Mysticism

IN THE SCIENTIFIC REVOLUTION

Reason, Experiment, and Mysticism

IN THE SCIENTIFIC REVOLUTION

M. L. Righini Bonelli
and
William R. Shea
Editors

ILLUSTRATED

SHP

SCIENCE HISTORY PUBLICATIONS
NEW YORK · 1975

First published in the United States by
SCIENCE HISTORY PUBLICATIONS
a division of
Neale Watson Academic Publications, Inc.
156 Fifth Avenue, New York 10010

First Edition 1975
(CIP Data on final page)
Designed and manufactured in the U.S.A.

Publisher's Note No effort has been made to superimpose a rigid uniformity of style or syntax. The authors' wishes have been followed wherever possible. Controversial matters have been resolved with the assistance of the editors.

If the publishers have unwittingly infringed the copyright in any illustration reproduced, they will gladly pay an appropriate fee on being satisfied as to the owner's title.

CONTENTS

Introduction: Trends in the Interpretation of Seventeenth Century Science

WILLIAM R. SHEA

The interplay and the natural influence of theory and experiment in the rise of the experimental method has often been discussed. What is novel, both in the history and the philosophy of science, is the growing interest in *mysticism*, a catch-all word used to characterize various forms of hermeticism, alchemy and Renaissance naturalism. Among these currents, which until recently could be described as "non-modern", hermeticism or the belief that there existed a secret tradition of knowledge that gave a truer insight into the basic forces of the universe has enjoyed particular favour. Less studied, but equally important, is the group of Paracelsian physicians and philosophers whose works are analysed by Allen G. Debus in this volume. But in both cases, we stand very much in need of an appraisal of those once famous men whose writings have sunk below the horizon of post-Newtonian science.

Mysticism, which looms so large today, was excluded from the history of science and the history of medicine on the grounds that it was irrelevant to the mainstream of scientific thought. Positivist historians of science directed their attention to that segment of the past that could be related with present concerns and contemporary methods. But as historians of science became more conscious of the need of acquiring a feeling for the contemporary resonance of ideas and concepts, they were led to see that events as experienced by actual men can only be understood by becoming involved in their general outlook. This meant that historians of science had to busy themselves with false starts as well as successful arrivals. The distinction between wrong and right steps, however, was assumed to be clear and to rest on the bedrock of scientific achievements. In recent years, doubt has been cast on the overall achievement of modern science and, consequently, the clarity of the distinction has been blurred. This had made it easier for historians of science to take a further step in reappraising the historiography of science and to raise the crucial question: Are "mystical" traditions and "modern" elements in the Scientific Revolution antagonistic or, paradoxically, complementary extremes?

It is usually said that two dominant strands made up the warp and woof of the new science. One expressed itself in the mechanical philosophy and had its roots in the atomism of the ancient world, the other was concerned with the exact mathematical description of phenomena and traced its origins to the Pythagorean tradition. This now seems an incomplete picture, for it is just as likely that two other ancient traditions contributed to shaping what we have come to know and to rely on as the experimental method. These are the Aristotelian philosophy with its logical rigour and its insistence on a realistic interpretation of nature and the hermetic tradition with its reliance on experiment, its appreciation of the crafts and its utilitarian outlook on science. We sometimes forget that latter-day opponents of Aristotle often retained more of his views than they would have been fond of admitting, and we have only too frequently read the history of mysticism in the works of those who condemned it. Until we learn to see Paracelsus, Bruno, Campanella and their kindred spirits as they saw themselves, we shall continue to be trapped in the prejudices of their adversaries.

This new research programme, so brilliantly illustrated in the articles by Debus and Westfall is, unfortunately, in danger of being jeopardized by those who present the issue as one of rational-vs-irrational elements in the development of modern science. This is profoundly misleading for the question is not whether men wished to be rational or irrational but whether the hermetic and mystical traditions had, in the seventeenth century, as good credentials of rationality (as it was then understood) as the mechanical philosophy. Debus and Westfall address themselves to this question with exemplary open-mindedness. This is no easy task because losers are always wrong, and we see ourselves as the heirs of the victorious mechanical philosophers. In retrospect, Fludd and van Helmont were in the dark about the best strategy to promote human understanding, but for their contemporaries, Mersenne, Gassendi and Kepler, they were dangerous rivals in the quest for the true philosophy of nature. It is hard for us to imagine that alchemy, which is now relegated to the lunatic fringe, was at one time a serious contender for the throne of reason. But until the correct route to scientific knowledge had been discovered (in Bacon's phrase "not by arguing but by trying") it was an open question what kind of world men found themselves inhabiting and radically different answers could claim a reasonable man's consideration.

Debus makes a major contribution towards a reassessment of the hermetic and mystical background to much of seventeenth century science. As with so many other things, it is largely a matter of perspective. When we look at the Paracelsian iatrochemists, who combined the study of medicine and chemistry, not, superficially, as self-deluded mystics, or, anachronistically, as chemists who failed to bring about a chemical revolution, but as the Chemical Philosophers they considered themselves, their programme becomes both more interesting and more reasonable. We begin to see why their appeal was so broad and why their ideas were discussed by people interested in religious and educational reform as well as by physicians and philosophers.

For the Chemical Philosophers, the unity and harmony of the universe was an obvious consequence of creation by a wise and benevolent God. This belief was shared by the vast majority of seventeenth century thinkers who instinctively shunned the materialistic implications of the mechanical philosophy. They could not fail to attend to a new philosophy that professed to achieve better scientific results with principles that conformed to Christian precepts rather than those of the "Pagan Aristotle". But this religious tone was not generally confessional, and many radicals welcomed the prospect of bypassing the authority of the established churches to pursue an independent study of the two God-given books: the Bible and the Created Book of Nature.

God created all things that they may be perfect and the iatrochemist considered it a duty as well as a privilege to be called upon to assist the natural process towards perfection by removing the pure from the impure and allowing the seeds of things to attain their ends. Baser metals could be hastened toward perfection and ill-tempered constitutions could be amended with the aid of chemistry. The spagiric art, as after Paracelsus alchemy was defined, worked for the well-being of nature whether vegetable, mineral or animal. It was a practical art and alchemists were not afraid of working with their own hands. They ridiculed the pedantry and logical exercises of the Schoolmen who cared for nothing but the words of Galen and Aristotle. The philosophy they appealed to was neither the qualitative physics of Aristotle nor the mathematical harmonies of the Neo-Platonists but Renaissance naturalism in which macrocosm-microcosm analogies provided clues and, in some cases such as Fludd, even evidence for the truth of general theories. This organic vision of the universe and the insistence on experiments made them scornful of mathematical rigour. When Fludd accused Kepler of belonging to the "vulgar crowd that pursues quantitative shadows", he was voicing the persuasion, common among alchemists, that mathematical reasoning, however coherent, rests on abstract assumptions that are not borne out in experience. They inveighed against the mechanical philosophy, not because it was a new form of knowledge, but because it appeared to be a reductionism founded on the purely speculative premise that everything in nature can be explained on the analogy of a clock.

Debus argues that van Helmont's chemical philosophy presented itself to the rational enquiring mind of the mid seventeenth century as a genuine alternative to Cartesian mechanism and that it reacted in a complicated way on this philosophy to bring about a more experimental approach to chemistry. This claim, which would have sounded preposterous twenty years ago, must now be taken as one of the most serious reassessments of the development of the scientific method. We are now apprised that the mechanical philosophy was often a thin veneer to cover a basically organic vision of the universe. In the case of Kenelm Digby, for instance, it was a thin sheet of ice that barely concealed the objectionable features of a philosophy that he was unable to forsake.

The shortcomings of the mechanical philosophy were particularly glaring in medicine and chemistry. The new conception of nature proclaimed the need of purging the body of natural philosophy of the occult and looked for the mechanical reality behind every phenomenon. Yet it offered no way of discriminating between competing invisible mechanisms and the unity of chemistry dissolved into as many theories as there were philosophers to imagine various hidden springs of matter. More fundamentally, still, the Cartesian programme broke down in its attempt to explain life as a mere appearance, a species of occult quality.

With the advantage of hindsight, we can see that the mechanical philosophy had to be given a broader base to include the chemical and biological data that it initially precluded itself from understanding. We are not faced with a simple clash between two completely contrasting viewpoints but with a complex interaction between two systems influencing each other and eventually giving rise to a new and richer method of scientific investigation.

If the contribution of van Helmont's iatrochemistry to the development of modern chemistry seems assured, it is more difficult to assess the importance of hermeticism in the genesis of Newtonian physics. In a critical description and chronological ordering of Newton's alchemical writings, R. S. Westfall offers striking evidence that Newton's interest in alchemy continued unabated between 1670 and 1696, the year he left Cambridge to become Warden of the Mint. The bulk of his notes is impressive (well over half a million words), but more revealing still is the fact that Newton attempted to write alchemical treatises himself, and that alchemy was his only abiding passion. He devoted merely two years, 1664–1665, to mathematics and from that time on would only turn to it when solicited. He concerned himself with optics for a brief period around 1670 but he never returned seriously to it again. Mechanics and dynamics held his attention for a short while in the 1660s and then only in the two and a half years that produced the *Principia.* It is hard to escape the conclusion that this great work, which we see as the culmination of Newton's career, may have seemed to him as an interruption or no more than a partial fulfilment of a much more grandiose plan.

Newton was convinced that there was an underlying unity to alchemy and that a comprehension of the alchemical "work" could be achieved both by comparing the various symbols they used and by making experiments. Although the records of Newton's chemical experiments are seldom couched in the imagery familiar to alchemical writers, Westfall believes they were performed against the background of two fundamental tenets of hermeticism: first, the belief that generation by male and female is a universal process at all levels of nature, and, second, the need to purge and cleanse in order that the spiritual seeds of things may reach maturity.

Now that we know how deeply fascinated Newton was by alchemical treatises and what store he put in them, we shall have to reconsider, as Westfall suggests, the route that led to the *Principia* themselves. In the standard accounts of the history of science, the law of universal gravitation is presented as the outcome of a fruitful suggestion by Robert Hooke and an opportune question by Edmund Halley. It is difficult to see how the "noble art" could have affected the course of rational mechanics, but this may be simply because we are not trained to look for likely evidence. The first step in the right direction is perhaps to don the thinking cap of the Cartesians and to learn to fear, with them, that Mr. Newton was introducing an occult quality under the guise of attraction. The next step is to ask whether Newton's disparagement of occult qualities in the famous Query 31 of the *Opticks* is more political than real, and whether it could have stemmed from a desire to repel the charge of obscurantism rather than from the fear of becoming an obscurantist.

Paolo Casini and Marie Boas Hall agree that Newton's radical revision of the prevailing mechanical philosophy of nature by embracing the notion of action at a distance no longer seems unconnected with his alchemical activities. They regard this view as promising but they urge that it be followed with caution. Once hermetic themes have been unearthed, there is a real danger of seeing them everywhere. If we are to make any headway in our appreciation of the influence of mysticism, we must be willing to individuate and specify its various manifestations. For instance, the pervasive "book of nature" analogy is in itself highly ambiguous. While it was at the tip of everyone's pen it did not carry the same meaning for every writer. Galileo and Kepler believed that God wrote the book of nature in mathematical characters but Bruno and Campanella interpreted it as the work of the Arch-Magician. It is true that mathematics can work wonders and that magicians juggle with numbers but the methodological programmes of the two schools were profoundly at variance. The "book of nature" metaphor used by all parties serves to conceal these differences. The following famous poem by Campanella could have been written by Galileo or Kepler but it would then have carried a different message.

> *Il mondo è il libro dove il Senno eterno*
> *scrisse i propri concetti, e vivo tempio*
> *dove, pingendo i gesti e 'l proprio esempio,*
> *di statue vive ornò l'imo e 'l superno;*
> *perch'ogni spirito qui l'arte e 'l governo*
> *leggere e contemplar, per non farsi empio,*
> *debba, e dir possa: "Io l'universo adempio,*
> *Dio contemplando a tutte cose interno."*
>
> *Ma noi, strette a' libri e tempii morti,*
> *copiati dal vivo con più errori,*
> *gli anteponghiamo a magistero tale.*
> *O pene, del fallir fatene accorti,*
> *liti, ignoranze, fatiche e dolori:*
> *deh torniamo, per Dio, all'originale!*

(The world is the book where eternal Wisdom wrote its own ideas, and the living temple where, depicting its own acts and likeness, it decorated the height and depth with living statues; so that every spirit to guard against profanity, should read and contemplate here art and government, and each should say: "I fill the universe, seeing God in all things."

But we, souls bound to books and dead temples, copied with many mistakes from the living, place these things before such instructions. O ills, quarrels, ignorance, labours, pains, make us aware of our falling away: o, let us, in God's name, return to the original! [Trans. by George Kay, *The Penguin Book of Italian Verse,* pp. 203–204]).

Kepler offers a particularly fascinating case study in the pitfalls of historical interpretation. He was a mystical visionary but he kept his ideas under the control of mathematical reasoning and physical experience. The student of Kepler should feel the intensity of his vision on his own pulse but also be prepared for a long process of thought to understand the sources of his vision and to be able to weigh its numerous elements. These must be studied in detail if their contribution to the whole is to be understood. Yet it would be wrong to assume that an exhaustive study of the component parts of his system is sufficient in itself. For instance, although Copernicus spoke of the sun as the "*gubernator*" of the planets, he did so only in the sense of one who is at the helm, not as the motor of the planetary revolutions. The first to suggest that the sun, because of its central position, is the source of the motion of the planets, was Rheticus in his *Narratio Prima.* But this idea was not taken up until Kepler saw the necessity of a physical explanation of the revolutions of the planets. The concept that the sun is the "*gubernator*" of the planets is Stoic in origin, and Kepler may be considered as having extended and mathematized this notion. The terminology he uses, "*anima movens*", indicates the source of his natural philosophy but it would be erroneous to reduce it to a modified form of an organic vision of the world. Blinkered accuracies about Kepler's vocabulary will tell us very little about his conception of nature.

Faced with a task of this kind it is tempting to take a direct line: to project into Kepler a set of opinions that can be seen to fit some area of his work and then view everything against them. This was done with literary panache by Koestler in *The Sleepwalkers* and the vividness of his description masks the one-sidedness of his interpretation. By reducing Kepler to a psychiatric case, Koestler has secured for him a twentieth century notoriety that he would not have had otherwise. Newton has yet to be made fashionable in this way and it is interesting that none of the contributors to this volume has chosen to make much of Frank Manuel's suggestion that Newton be, in a sense, psychoanalyzed.

They have preferred to tackle a different set of questions and to ponder the problem of Newton's intentions. Was Newton reading the hermetic literature to find *evidence* or merely *phenomena* about the physical world? It is along these lines that a fruitful debate is now open.

Paolo Rossi, whose *Francis Bacon, From Magic to Science* was a pioneering work in the rediscovery of the "mystical" roots of the Scientific Revolution, is concerned lest, in our concentration on the roots, we forget the characteristic features of the full-grown tree. The seventeenth century was a distinctive age in that it witnessed an enormous expansion of knowledge. Overseas expeditions revealed the existence of new plants and new animals. Intensified anatomical research uncovered new information about what had been considered well known, the circulation of the blood was discovered, the theory of spontaneous generation was refuted, and new evidence about the age of the earth ushered in a new way of thinking about Adam. New instruments opened up new fields of investigation. The telescope revolutionized the study of the heavens, the microscope disclosed new realms of life, the airpump made a virtual vacuum possible, the barometer and the thermometer brought atmospheric pressure within the bounds of measurement, and the first precision clock enabled man to speak of time in a new idiom.

But the Scientific Revolution was primarily a matter not of new facts, but of new ways of looking at old facts. Galileo may have been the first to turn the telescope to the skies but the heliocentric theory he embraced had been elaborated without its aid and Kepler had worked out the correct path of Mars before he heard of its invention. The microscope was not used systematically before Malpighi. Harvey, whose work is largely posterior to the invention of this instrument failed to make use of it. It is not that Harvey rejected the microscope as technically inadequate for he occasionally used lenses of low magnification. It is simply a problem of method: the necessity of considering the existence of entities invisible to the naked eye though in principle observable did not enter into his research programme.

The abstract nature of the models used to understand nature and their extension to the realm of man (illustrated and discussed in the articles by Belloni and Duchesneau) can be ascribed to the mechanical philosophy. So can the distinction between primary and secondary qualities, the stress on the "subjectivity" of sensory experience and the "objectivity" of mechanical laws, as well as the idea of God as a watchmaker. But many of the characteristic features of the new scientific spirit owe at least as much to the "mystical" traditions. The new appreciation of the crafts and the need for experiment were long recognized by iatrochemists. The same can be said of the ideal of scientific co-operation and the notion of progress as growth of mankind. The infinity of the universe and the rise of cultural relativism were mainly discussed by hermeticists such as Bruno and Campanella. Even what we refer to as Baconian utilitarianism or the pursuit of science for the relief of man's estate was a frequent theme among the exponents of the Chemical Philosophy. To speak of this idea as typically Baconian is merely to vouch for the literary gifts which have continued to make him read when his less fortunate contemporaries have sunk into oblivion.

Equally new is the relationship of "art" to "nature" and the ideal of knowledge as "construction". The nominalistic notion that we can only know what we can make runs through the works of Mersenne, Gassendi and Hobbes. This view precludes any "deep" knowledge of nature, but it is a philosophical stance that is not shared by the main artisans of the Scientific Revolution who were thoroughgoing realists. Mersenne and Gassendi argued in innumerable folios that if we can only know what we can construct, then human understanding is precious little since God alone can create nature. Galileo, Kepler, and Descartes, on the contrary, insisted that God was a geometrician in his creative labours and that mathematics laid bare the foundations of the cosmic structure. In his *Mysterium Cosmographicum,* Kepler is conscious of speaking paradoxically when he says that God created "nature" according to the "art" of an architect. It is tempting to suggest that those for whom the new philosophy (however superior to the old) gave no real insight into the recesses of nature were men who had once shared the Aristotelian persuasion that true understanding is abstraction of "essences" and "substantial forms". When they rejected these "forms" as a scientific delusion, they proceeded, with the enthusiasm of neophytes, to proclaim the bankruptcy of human knowledge itself. But for those men, like Galileo and Kepler, who were familiar with the tenets of the old school but never belonged to it, matters were different. When they came to see that "essential forms" were mere abstract notions, they asserted that the true world was composed of atoms that move according to intelligible mathematical laws. Mersenne and Gassendi were, in a sense, reluctant converts to a new way of viewing nature; Galileo and Kepler travelled farther along the road that led away from Aristotelianism. This is not to affirm that they were completely divorced from the past but simply that it would be erroneous to read their works as if they were merely refurbishing old ideas.

For when all is said and done there was a radical change in the seventeenth century. The Scientific Revolution consecrated a new method that was slowly to transform a civilization organized around Christianity into one centred on science. This way of thinking is usually described by the short-hand term of *the experimental method,* meaning the active questioning of nature under conditions defined by the experimenter rather than the mere observation of the phenomena that spontaneously present themselves. Of course, there is no one rigid experimental method that is uniformly applied throughout all the sciences, and even the outline of what is now called the hypothetico-deductive method was shown by A. C. Crombie in his *Robert Grosseteste and the Origins of Experimental Science* to have been discussed by the Bishop of Lincoln and his disciples in the Middle Ages. But as Crombie himself perceptively notes, one should not equate philosophical discussions of method with actual practice of scientists, a word of caution, let it be said, that is not always heeded by those who are mainly interested in logical and epistemological theories about science.

Empirical investigation, in the sense of the formulation of hypotheses on the basis of empirical data stated with sufficient rigour to allow the hypotheses to be stringently tested, was not widespread before the seventeenth century. This method stands in constrast not so much with the Chemical as with the Aristotelian philosophy. However at odds the new scientists may have been on points of philosophy, they rose to a man when it came to pass judgment on the method of the old physics which they considered a method of exposition not a method of discovery or proof.

We can hardly overestimate the importance of the ideas that a scientist brings to his scientific work especially those which concern what he is looking for and what constitutes evidence for a theory. Now what counted as evidence for the old physics (which borrowed its paradigm of scientific reasoning from the *Posterior Analytics*) seemed to the new generation to be completely at variance with what ought to be treated as evidence.

Aristotle's ideas about what it means to know — to know scientifically — were so diffuse that they must be recalled here, however briefly, if the novelty of the new experimental method is to be appreciated.

Aristotle's views on scientific knowledge rest on a cosmological assumption about the nature of the universe and an epistemological assumption about the nature of the human mind. The first assumption is that there exists a world of essences arranged in hierarchical structures that can be laid out logically. This world is basically of a teleological sort and essences are disclosed by their finalities. The second assumption is that the human mind is gifted with the quality of insight which enables it to grasp essences, to "see" the universal in the particular. This power of abstraction, as it is sometimes called, allows man to perceive the premises of natural science as necessary truths.

From these two general assumptions there follows a method of knowing. Since concepts are derived immediately from ordinary experience, what characterizes a body is its habitual mode of behaviour, some "natural motion" which expresses its "form". Earth, for instance, naturally moves downwards because in its normal context it will either be at rest or falling towards the centre of the world. Everyday language is sufficient to state the nature of the elements which are defined by the sensory qualities of hot and cold, dry and moist. "Essences" once named are located in a network of concepts that are related through mutual definitions. There is no need for technical concepts such as those of mathematics and there is no question of a quantitative structuring of experience. Quantity is irrelevant because it can only tell how much of a substance we have, not what kind.

In this perspective, science was basically a descriptive enterprise. Aristotle objected to Plato's assertion in the *Timaeus* that the world is composed of atoms of certain shapes because the warrant he offered came not from observation but from pure geometry. Neither could Aristotle entertain the idea that a body of scientific doctrine was a theory in the modern sense of a logical and mathematical network of laws that have to be anchored in the empirical data available. His teleological bias meant that nature was considered as a process and this turned speculation away from questions of structure and mechanism towards questions of function and development.

Sciences, such as astronomy, that use mathematics as their syntax, were not, strictly speaking, physical sciences. They were said to move on a level of abstraction that did not consider physical entities in their entirety. Such an abstraction gave no direct insight into the nature of things and remained hypothetical. Epicycles and the like were calculating devices not real explanations of the motion of the planets.

Since nature was to be discovered not by asking what a thing would do in every conceivable context, but what it does in its natural context, it comes as no surprise that experiment is not even discussed in the *Posterior Analytics.* The artificial simplification of the natural context would have seemed wrong-headed.

Modern science, on the contrary, assumes that it is by imposing on nature a set of artificial conditions dictated by a precise question that nature will reveal its secrets. For instance, Torricelli would never have filled a glass tube with mercury and erected it on a dish if he had not had a specific design in mind. Nature only speaks in the language in which it is interrogated.

The development of this modern method of empirical investigation with its twin aspects of *idealization* and *contrivance* was a slow process. However hostile to the Aristotelian establishment, Galileo, for one, remained profoundly influenced by the model of scientific explanation in the *Posterior Analytics* and his lifelong passion was to show that the Copernican system could be demonstrated with all the rigour demanded by the Aristotelian canons. In his important essay in this volume, A. C. Crombie conclusively establishes that Galileo was well read in the best scholastic literature of his day and that his early writings owe much to three Jesuit professors of the Collegio Romano. Crombie points out that Galileo's running argument against Aristotle can obscure the fact that he never wavered in his Aristotelian belief that there is a literally and uniquely true physical theory, that it can be discovered and that alternative theories are consequently false. Where Galileo broke with the Master was in his conception of the nature of this physical reality. He sought to transcend the limitations of Aristotelian empiricism by claiming that reality was mathematical in form, and that mathematical theory should determine the very structure of experimental research. In this he was following Archimedes, who was regarded as a Platonist by the doxological tradition. Galileo's mathematical essentialism, in fact, must be interpreted against the background of the Platonic revival in Italy, but he differed from Plato precisely in his essentialism. Plato maintained that the physical world was a copy or likeness of a transcendent world of ideal mathematical forms, but it was an imperfect copy and so physics was not absolute truth but an approximation. Galileo, however, held that the world actually consisted of the mathematical primary and secondary qualities and their laws, and that these laws could be known with absolute certainty.

But what is crucial is that Galileo's actual procedure ran counter to Aristotelian methodology and that he was, in practice, undoing what he strove so hard to achieve by developing an approach to nature that precipitated the downfall of the Aristotelian ideal of scientific knowledge. In the long run, causes were superseded by laws, and knowledge of essences by determinations of regularities, even if the process was neither simple nor instantaneous.

In his clear and deft analysis of Galileo's new science of motion, Stillman Drake corrects two pervasive mistakes about Galileo's method, namely the assumption that he started from the so-called Merton Rule of mean speed, and the common belief that the impetus theory resisted mathematization. One of the highlights of Drake's paper is the discussion of Honoré Fabri's claim that his Aristotelian theory was better because it provided a cause for acceleration while Galileo could not.

Guglielmo Righini's article on Galileo's lunar observations is a timely reminder of the danger of judging astronomical observations without ascertaining with care the date on which they were made. Galileo has been roundly condemned as a second-rate observer whose perception was blurred by the excitement of discovery when, on closer inspection, it can be shown that he was remarkably faithful in his description of the main features of the moon. The indictment of Galileo as a poor experimentalist because of his diagrams of the lunar landscape turns out to have been a hasty generalization, made for philosophical rather than historical reasons.

The challenging essays in this volume are reports on work in progress and it is too early to tell what the complete picture of the method of seventeenth century science will look like. It is now clear, however, that hermeticism and alchemy made a positive contribution to the experimental method by stressing the importance of observations, freeing science from the weight of inherited authority, as well as by recognizing the value and dignity of the crafts, and emphasizing the utilitarian goal of scientific knowledge. The new method was not elaborated in one day and unresolved tensions are everywhere apparent. Kepler often speaks as though mathematics provides not only the *language* of science but the *evidence* of science, and Galileo never freed himself completely from the Aristotelian ideal of a certain and exhaustive knowledge based on the comprehension of the true cause of phenomena. The process was not completed until the end of the seventeenth century, but the gradual and often chaotic transformation of what constitutes scientific knowledge should not make us blind to the fact that we owe to this golden age our new outlook on the world.

The true Chemical Philosopher worships his God through prayer and the study of his written word as well as through the study of his Creation, Nature, in the chemical laboratory. From Heinrich Khunrath, *Amphitheatrum Sapientiae, solius verae, Christiano-Kabalisticum, Divino-Magicum, nec non Physico-Chymicum, Tertriunum, Catholicon . . .* (Hanoviae: Gulielmus Antonius, 1609), folding plate at end.

The Chemical Debates of the Seventeenth Century: The Reaction to Robert Fludd and Jean Baptiste van Helmont

ALLEN G. DEBUS

A considerable body of recent research both in the history of science and the history of medicine has concentrated on what may conveniently be termed "non-modern" themes in the period of the Scientific Revolution. In this work it is perhaps Hermeticism that has emerged as the subject of greatest interest. Less thoroughly explored has been the work of the Paracelsians whose combined interest in chemistry and medicine sets them apart from other Hermeticists. The work of these authors has generally not seemed to be in the mainstream of the history of medicine to medical historians—nor in the mainstream of the history of science to scientific historians. Yet, the Paracelsian debates of the late sixteenth and the seventeenth centuries were widely broadcast at the time and it is clear that they were a subject of great concern to physicians and natural philosophers alike. These debates touched upon a broad spectrum of subjects that range from religious and educational reform on the one hand to scientific methodology and practical medicine on the other. Nor were these debates limited in scope to a handful of dissident radical elements. Rather, they excited the interest and concern of some of the major figures of the period.

Although I believe that an assessment of the work of the Paracelsian and the chemical physicians as a whole is badly needed for our understanding of the Scientific Revolution, this will not be the purpose of the present paper. There are still too few engaged in research on the medico-chemical texts of this period, and we are still at a stage where additional descriptive accounts of the literature are required. In the limited space available, it will be our goal to give some account of the scope and issues of this debate by focusing on the work of two of these Chemical Philosophers — Robert Fludd (1574 – 1637) and Jean Baptiste van Helmont (1579–1644)— and, more particularly, on the reaction to their work by the European community of scholars.

The Chemical Philosophy

Prior to discussing the controversies that surrounded Fludd and van Helmont we should briefly note the background to these debates and give some indication of the development and content of the philosophy of nature subscribed to by most late sixteenth and early seventeenth century iatrochemists.

It is best to think of the Parcelsians as Chemical Philosophers — a term they used with regularity themselves. However, if we are to use it as well, we must not limit ourselves to a twentieth century definition of chemistry. In the sixteenth and seventeenth centuries chemistry was most commonly defined as the spagyric art, an art in which the subject matter was primarily the separation of the pure from the impure.[1] It was also an art dedicated to the perfection of the imperfect in nature. Thus, the imperfect metals might surely be hastened on their slow path to nobility through the skill of the alchemist. Similarly, the ills of man might well be treated by this art. The alchemist was both a physician of man and nature. That Scriptural authority might be found for the semi-divine nature of the physician assured the frequent citation of *Ecclesiasticus* 38.[2] In addition to these definitions, however, the sixteenth century chemist thought of himself as a natural philosopher — one to whom the secrets of nature would be unlocked through the key of chemistry.[3]

This Chemical Philosophy was to be a universal philosophy of nature founded on new observations and indisputable philosophical precepts which conformed to religious truth. The teachings of the schools were rejected as moribund and impious. Beyond this, these authors complained that the universities were dominated by logic. Clearly related to this was their distrust of mathematics as an interpreter of nature.[4] With some authors we find the term practically equated with logic. We find Galen and Aristotle faulted time and time again for their reliance on a mathematically oriented medicine and philosophy of nature. In contrast, the Paracelsians argued, the true natural philosopher must be guided principally by two books, the written record of Holy Scripture and the Created book of Nature. If the first was to be read as a book, the second was to be studied through the chemist's own observations and analyses.[5] This plea for reform stands out in the work of Paracelsus. Idle speculations and theories are to be rejected in the study of nature. In contrast to the Galenist, the Paracelsian chemical physician is to be a man who is not afraid to work with his own hands, a pious scholar who praises God in his work and who lays aside all the vanities of his tradition-bound competitor. His delight is found in a knowledge of the fire while he learns the degrees of the science of alchemy.[6] In his *Labyrinthum medicorum* (1538) Paracelsus referred to no printed books. Instead we are told to seek out God in nature. We must turn to the book of the heavens, the book of the elements, the book of man, the book of alchemy and the book of medicine. In effect, printed books are useless for the reform of learning since this must be based on experience and observation. "*Scientia enim est experientia.*"[7] This was to be an undeniable maxim of the Chemical Philosophy, one which was repeated time and time again throughout the sixteenth and seventeenth centuries.

One also meets with an insistence on the truth of the macrocosm-microcosm analogy in the works of these authors. The universe was pictured as small and closely interconnected. Macrocosmic events affected man, and man as the microcosm could, in turn, affect the great world.[8] The earth itself was thought to be a living organism analogous to man in this vast and all-encompassing vitalistic universe.[9] The Creation account in *Genesis* was interpreted as a chemical process of separation[10] and it seemed only right to go beyond this to conclude that all nature — the macrocosm as well as the microcosm — must continue to operate chemically. One result of this was that physiological processes were to be interpreted chemically by the followers of Paracelsus.[11]

The rejection of the ancients — coupled with the insistence on a chemical interpretation of the first chapter of *Genesis* — permitted the Paracelsians to question the Aristotelian elements: earth, water, air and fire. For these they sometimes added and sometimes substituted a triad: Salt, Sulphur and Mercury.[12] The related four humors were also brought under heavy fire. Rather than an imbalance of these fluids affecting the body as a whole, disease was for the Paracelsians a localized abnormality within the body. And since this process was for them one that might be interpreted chemically, it might well be treated with chemicals. As for the theory of cure, the Paracelsians disagreed sharply with the Galenic belief that contraries cure. Rather, they turned to Germanic folk tradition which indicated that like cures like. In practice this meant that a poison in the body should be cured with a poison — that is, if it had been treated chemically to remove its toxic nature.[13] The theory of cure was thus closely linked with more general concepts: sympathetic action in nature and the search for universal harmony.

It would be erroneous to imply that this short *primer* is a complete one. Admittedly I have oversimplified a very complex subject. However, I hope to have sufficiently indicated that the Chemical Philosophers felt that they had rejected the "irreligious" natural philosophy and medicine of the ancients which they hoped to replace with a new system of nature and man — the key to which would be chemistry. It is possible and quite valid to show that many of their most cherished concepts were little removed from Aristotelian and Galenic explanations, but it is a fact that it was most common at the time for their enemies to view the chemists as a dangerous group of innovators whose goal was to overturn the entire body of respectable knowledge. In truth there did exist basic differences between the Galenists and the Paracelsians on crucial points and it would be difficult to ignore the sharp contemporary reaction that occurred. The stage was clearly set for a confrontation between the medical and educational establishments on the one hand and the Chemical Philosophers on the other.

The Controversy over the Fluddean Philosophy

Paracelsus had died in 1541 at the early age of forty-eight and there had been little reaction against his work during his lifetime. It was during the course of the second half of the sixteenth century that the Paracelsian debates began — and steadily accelerated. A widespread belief that Paracelsus had been able to cure hitherto incurable diseases resulted in a search for his manuscripts which were then scattered throughout Germany and Switzerland. These were published — often with lengthy commentaries — and within a few years there was enough of this literature in print for syntheses of Paracelsian dogma to be prepared by a number of authors. The first comprehensive attack on the system was that of Thomas Erastus (1572–1574) and in 1579 there was a highly emotional legal encounter between a Paracelsian court physician and the Galenist establishment in Paris.[14]

Although the Galenists were to triumph over their opponent in 1579, the results were far less decisive a quarter century later. The publication of Joseph Duchesne's *De priscorum philosophorum verae medicinae materia* . . . in 1603 was again to pit the medical establishment against the Chemical Physicians. This debate was to touch on all areas of this conflict from the problem of the Creation and the elements to the use of metallic remedies. Furthermore, numerous Latin editions and translations into the vernacular of many of the polemical tracts written in the period 1603 – 1609 made the various issues of this debate well known throughout Europe.

In the wake of this Parisian confrontation was the appearance of two short pamphlets in 1614 and 1615, the *Fama Fraternitatis* and the *Confessio* which were alleged to have been written by a group calling themselves "Rosicrucians."[15] In these texts there was presented a call for religious and educational reform. The former was to be fundamentalist and Protestant, the latter was to be largely Paracelsian and medical in tone. Far from impressive from a twentieth century vantage point, these works elicited an immediate response from contemporary readers. Several hundred tracts — both pro and con — are recorded as having been printed within a ten year period. Among these were works by two prominent physicians, Andreas Libavius (1540– 1616) and Robert Fludd.

Libavius perhaps may best be characterized as an anti-Paracelsian iatrochemist having a strong distaste for any mystical interpretation of natural phenomena. Although convinced of the practical importance of chemicals in medicine, he wanted no part of a Paracelsian chemical philosophy interpreting the universe in terms of the macrocosm-microcosm analogy. He had supported the Parisian chemical physicians against their hardline Galenist opponents in 1606, but nine years later he was to reject the Rosicrucian texts no less decisively. These seemed to be both mystical and Paracelsian to him — and even more damning, they promised the destruction of all aspects of ancient learning.

The attack by Libavius on the Rosicrucians was to be the cause for the first publication by Robert Fludd—already at this time a noted London physician and Fellow of the College of Physicians.[16] Hoping to be contacted by the members of the elusive order, Fludd penned a reply to Libavius (1616) in which he attacked the study of the ancients in the universities and called for a new learning based upon religious truths.[17] Arguing that there had been a decline in true knowledge since the time of Moses, Fludd suggested that in place of Aristotle and Galen the schools should turn to truths to be found in the study of alchemy, natural magic and a new medicine. He methodically criticized the liberal arts and specifically rejected the emphasis on logic to be found in the scholastic curriculum. This, he felt, was reflected in the academic approach to mathematics which was based upon definitions, principles and discussions of theoretical operations. Rather, Fludd wrote, the mathematician should turn to the mystic teachings of the Pythagoreans who had reached a certainty of belief in God through their study of numbers and their ratios. In this way we would be led to the concept of universal harmonies and to the very fabric of the world.

In this "Apology" for the Rosicrucians Fludd insisted that we must proceed to a new learning with a definite plan in mind. Accordingly, he listed a series of key questions which should form the basis of future inquiries.[18] We must, he wrote, consider the act of Creation and the formation and meaning of the elements. We must especially concern ourselves with the part played in Creation by the divine light of the Lord. This is no less than that vital spirit required for all life and motion. We must concern ourselves with all aspects of its action and pay attention to other concepts of interest — such as the views on atomism proposed by Democritus.[19] And as we turn our attention from the macrocosm to the microcosm, Fludd wrote that we must focus our attention on the assimilation of this vital spirit in the body.[20] Here he emphasized that this spirit was to be found in the air and that it entered our bodies through inspiration. The need to determine just how the spirit nourished our bodies would make necessary a new study of the body itself. We must determine how the spirit is separated from the gross air, and how it is dispersed throughout the body through the arterial and venous systems.

A second edition of the *Apologia* and the first volume of Fludd's long delayed history of the macrocosm and the microcosm were published in 1617.[21] The first was surely among the most important of the "Rosicrucian" texts and the second may be the most detailed exposition of the macrocosm-microcosm universe. Surely both works attracted considerable attention from a group of scholars who are today far better known than Fludd. For this reason — and also because a number of significant questions are involved—the reaction to Fludd's work is of interest for our understanding of the course of the Scientific Revolution.

The first major reply to Fludd's works was that of Johannes Kepler. In an appendix to the *Harmonices mundi* (1619) and in a later response to a reply by Fludd (1622) Kepler examined the English physician's use of mathematics.[22] On the surface this appears to be a clear-cut case of the "scientist" versus the "mystic." As Kepler presented his case, the differences between his own approach and that of Fludd were simple. His description of universal harmony might be called "mathematical" while Fludd's explanations were *"aenigmaticos, pictos, Hermeticos."*[23] The latter's works abounded with many symbolic pictures while his own instructed with true mathematical diagrams. Fludd delighted in shadowy aenigmas while Kepler sought to rescue the same phenomena from darkness and to bring them into the light.[24] Fludd borrowed fables from the ancients

Robert Fludd (1574-1637). From the *Integrum Morborum Mysterium: sive Medicinae Catholicae Tomi Primi Tractatus Secundus* . . . (Frankfurt: Fitzer, 1631), sig.: (1ᵛ. *From the collection of the author.*)

while Kepler built upon the fundamentals of nature with mathematical certitude.[25] The former confused things that he did not properly understand while the latter proceeded in an orderly fashion corresponding to the laws of nature. Again and again, Kepler returned to the sharp distinction to be made between the mathematician on the one hand and the chemist, the Hermeticist and the Paracelsian on the other.

And yet, as Pauli has indicated, the debate was not this simple. Kepler's development of the laws of planetary motion was the result of an interest that was deeply rooted in the search for mathematical perfection that forms part of the Pythagorean tradition. Convinced of the truth of the music of the spheres, he sought a movement of the planets in the same proportions that appear in the harmonious sounds of tones and regular polyhedra. No less than Fludd did he believe in an *anima mundi,* no less than Fludd did he argue for a near divine sun that belonged in the center of the world, and no less than Fludd did he believe in the stars as living entities.

Nevertheless, although we may point to real similarities, it is also true that the meaning of mathematics for Kepler was something quite different than it was for Fludd. The latter sought mysteries in symbols according to a preconceived belief in a cosmic plan. His proportions and harmonies were forced to fit these symbols. Kepler, perhaps just as obsessed with his symbolic spherical picture of the world, insisted that his hypotheses be founded on quantitative, mathematically demonstrable premises. If an hypothesis could not accomodate his observations, Kepler was willing to alter it. These two views were so opposed that the two men could not really understand each other. Fludd felt that he could honestly call Kepler one of the worst sort of mathematicians, one of the vulgar crowd who "concern themselves with quantitative shadows." In contrast, "the alchemists and Hermetic philosophers . . . understand the true core of the natural bodies."[26]

Although the Fludd-Kepler exchange is of considerable interest, the scope of the reaction to Fludd's publications among French scholars was to be much broader. In France the Paracelsian medical debates of the late sixteenth and early seventeenth centuries had been accompanied by a steady flow of new and reprinted chemical texts. In addition, an announced "visitation" of the Rosicrucians to Paris in 1623[27] was followed in August of the ensuing year with a debate at the residence of an influential Hermeticist, François de Soucy.[28] Here fourteen alchemical theses were defended: among them a denial of the *materia prima* and the doctrines of substantial form and privation. The sublunary world was affirmed to be composed of the two elements, earth and water plus the three Paracelsian principles. Other theses emphasized the world soul, various aspects of the specific elements and the atomic nature of matter. The actual meeting was dispersed on the order of the Parlement of Paris and a principal participant in the discussions, Estienne de Clave, was arrested. By the sixth of September the Doctors of the Sorbonne had officially condemned the theses and before the end of the year Jean-Baptiste Morin had published his own *Réfutation* . . . .

Clearly concerned by what appeared to be an ever-increasing interest in Rosicrucian and alchemical mysticism, Marin Mersenne (1588–1648) examined these subjects both in his *Quaestiones celeberrimae in Genesim* (1623) and his *La Verité des Sciences* (1625). It was particularly in the first of these that Mersenne attacked Fludd by name, but both works indicate the alarm with which Mersenne viewed the type of Chemical Philosophy subscribed to by the English physician. This is most evident in *La Verité des Sciences* where once again mathematics was proposed as the basis of a new science. As in the case of Kepler, Mersenne argued for a far different use of mathematics than did Fludd.

It would seem that Mersenne felt that a truly mathematically oriented science must first overcome the claims of the Chemical Philosophers. In some detail he discussed these claims in dialogue form between an alchemist, a sceptic and a "Christian Philosopher." For the alchemist no science was more certain than his own because alchemy taught through experience.[29] And while ready to agree that charlatans existed, this chemist argued that the emphasis placed upon observation in this field of knowledge had resulted in a new and significant system of elements and principles. There seemed little doubt that the works of Aristotle — admittedly filled with dangerous theological views — had been replaced by an approach more sound.

Mersenne's rejection of the views of the alchemist was firm. For the "Christian Philosopher" the recent condemnation by the Sorbonne had been just. These learned Doctors had rightly questioned the theological implications of the alchemical theses. Among them the alchemical espousal of atomism was noted as a position which might easily be overturned.[30] As for the alchemists' much vaunted "observationally" based system of elements and principles, Mersenne replied that the Paracelsian principles may yet be decomposed artificially.[31] Should this occur, these principles need no longer be considered to be elementary in nature.

And yet, Mersenne continued, if alchemy may be faulted on some points, it must not be rejected *in toto*. Rather, some method of control must be sought to avoid the dangerous pitfalls into which alchemists had fallen too often in the past. Mersenne suggested the establishment of alchemical academies in each kingdom which might take as their goal the improvement of the health of mankind. These academies might police the field not only by punishing charlatans, but also by actively engaging themselves in the reform of science. Allegorical and enigmatic terms such as "Christiano-cabalistique, Divino-magique, Physico-chymique," and the like would be discarded while in their place there would be substituted a clear terminology based upon chemical operations performed in the laboratory.

For Mersenne a reformed alchemy would steer clear of religious, philosophical and theological questions which were of absolutely no concern to it.[32] There were some for whom the subject seemed to serve as a counter-church and who argued that the most ancient theology, magic, and pagan fables were best explained through this science. Many, indeed, held to the chemical interpretation of the Creation. These dreams and speculation must be rejected at once if the subject was to gain the approval of the Catholic Church.[33]

Mersenne had referred to a number of chemical authors in both works, but in the *Questions on Genesis* he had singled out Fludd as a *"cacomagus, foetidae et horrendae magiae doctor et propagator, haereticomagus, brevibus submergendum fluctibus aeternis."*[34] Deeply wounded, Fludd replied to the French author in two works which restated his position on the Chemical Philosophy.[35] Here he described once more the analogy of the macrocosm and the microcosm, the harmony of the two worlds, the significance of the vital spirit and its dispersal through the arterial system. Ture alchemy, Fludd insisted, has as its goal the establishment of the entire Chemical Philosophy as a basis of explanation for both man and the universe.

It is clear that Fludd's understanding of an *"alchemia vera"* was precisely what Mersenne sought to avoid. Above all, Fludd was disturbed by Mersenne's warning that alchemists should dissociate themselves from religious matters. On the contrary, he felt that chemists and theologians have a common subject to investigate. It is a part of practi-

P. Dupin Sculp
MARIN MERSENNE
Religieux de l'Ordre des Minimes Thologi
Philosophe et Mathematicien celebre né à
Oyse au Maine Mort à Paris 1648 âgé de 60 ans

Portrait of Marin Mersenne (1648) from *Correspondance du P. Marin Mersenne Religieux Minime,* Publiée et anotée par Cornélis De Waard (vol. 7, Paris: Édition de Centre National de la Recherche Scientifique, 1962), frontispiece.

cal Theology which "we think to be nothing else but mystic and occult Chemistry."[36] This subject is one that seeks to comprehend the Creation and the spirit of life. Nature and supernature are clearly united — and chemistry serves as a key to both.

In despair Mersenne sent a collection of Fludd's works to his friend Pierre Gassendi with an appeal for aid late in 1628. In little more than two months the latter had completed his critique. Predictably, Gassendi rejected Fludd's explanation of the elementary principles and the Chemical Creation. Gassendi noted Fludd's views on atomism, but when confronted with Fludd's rejection of Copernicus and Gilbert, he could only conclude that "he understands another non-volatile earth and central sun than that commonly understood by us." When Gassendi discussed the distinction made by Fludd between a true and a false alchemy, he fully supported the views of Mersenne. Here, in an impassioned passage, Gassendi complained of an interpretation that would make "alchemy the sole Religion, the Alchemist the sole Religious person, and the tyrocinium of Alchemy the sole Catechism of the Faith."[37]

Gassendi's analysis of Fludd's philosophy is of added interest because of a lengthy discussion of the cardiovascular system. Fludd had earlier (1623) described the circulation of the vital spirit in the arteries in terms of the macrocosm-microcosm analogy.[38] Mersenne, already aware of William Harvey's *De motu cordis* by the end of 1628, had sent it to Gassendi along with the works of Fludd — surely believing its author to be a disciple of his adversary. After reading both authors Gassendi was able to clearly distinguish between the experimental work of the one and the "mystical anatomy" of the other. But although he preferred Harvey to Fludd, Gassendi rejected both since he firmly believed in the existence of the interventricular pores in the septum of the heart. Since they exist, he argued, "they ought not to be useless, and there is a purpose in readiness, the percolation of the blood out of the right vessel into the left; and I might argue that the arterial blood is derived in this manner."[39] Gassendi thus rejected the concept of circulation in 1629.

Fludd's reply to Gassendi, printed in 1633, was to return to the same subject. Gassendi's belief in the interventricular pores had been based upon a single dissection he had witnessed as a student. The situation was far different in the case of Fludd who had observed Harvey search repeatedly for these pores. "Not in any one of the many cadavers that he examined did he find such a septum; and neither I nor any others who with most acute and almost lynx-like eyes saw this when we examined the septum of the heart." A single example simply would not suffice to prove Gassendi's view. Thus, on this point the mystical alchemist informed the observationalist that "we know and speak as experts" when we "assert with confidence that the septum of the heart is not ordained by nature to that purpose called for by Gassendi."[40]

It is not necessary to discuss the remaining years of this debate in any detail. This may best be followed through Mersenne's correspondence. Some thirty-five extant letters from the period 1631 – 1633 attest to a very real contemporary interest throughout Europe. Gassendi himself was not to reply to Fludd's *Clavis* of 1633, but Jean Durelle did prepare an answer in 1636.[41] Fludd's death the following year was not to still the debate since his *Philosophia Moysaica* appeared posthumously in 1638. Mersenne's continued concern may be followed through a number of letters from the years 1640 – 1642. Here he indicated his esteem for the work by Durelle while pressing Theodore Haack and John Pell in London for an English edition to which he promised to add an appendix of his own refuting Fludd's views on the weapon-salve. Clearly Fludd's death had done little to lessen Mersenne's concern over the possible influence of the English physician's dream of a "new philosophy" which was so opposed to his own.

Jean Baptiste van Helmont (1579–1644) and the Chemical Philosophy

In his search for support against Robert Fludd, Mersenne had contacted many European scholars. Among them was Jean Baptiste van Helmont[42] who was to carry on a rich correspondence with the French *savant.* Of this some fourteen letters survive from the years 1630–1631. In one of the earliest of these (19 Dec. 1630) van Helmont answered a query on the value of Gassendi's then recent reply to Fludd. The Belgian physician-chemist was unequivocal as he referred to the *"fluctuantem Fluddum"* who was a poor physician and a worse alchemist — a superficially learned man on whom Gassendi should not waste his time.[43]

It is of some interest to note that van Helmont — at least temporarily — may have been included within the Mersenne circle. In fact he, too, was a Chemical Philosopher and much of his work was characterized by concepts and attitudes already condemned in Mersenne's *La Verité des Sciences.* For this reason it is of considerable significance to examine the views of van Helmont. Here, in contrast to Fludd, few of whose works were republished after his death, we find an author of great influence whose restatement of the Chemical Philosophy was to become the basis of the iatrochemical school of the seventeenth century.

The early ages of van Helmont's *Ortus medicine* contain autobiographical material. Here we read of a tormented young scholar who found his training at the University of Louvain useless. He refused the degree, Master of Arts, and "seeking truth, and knowledge, but not their appearance, [I] withdrew my self from the Schooles."[44] Not long after

J. B. van Helmont, portrait from his *Aufgang der Artzney Kunst* (Sulzbach, 1683). *Courtesy of the Department of Special Collections, Library, University of Chicago.*

this he returned again to academic life only to find magic, Stoicism and medicine also valueless in his quest for certainty. Now, older and wiser if no less disillusioned, he accepted the degree Doctor of Medicine (1599). We find him in the early years of the new century in England and France mingling easily in court circles. Noted by many for his brilliance, he found himself receiving a number of tempting offers — all of which he declined. Still deeply troubled, he married in 1609 and retired with his bride to his home at Vilvord to devote himself to the study of nature. It was in the following years that van Helmont became deeply immersed in the works of Paracelsus and his disciples.

Van Helmont's search for truth was a personal one. But as the Rosicrucian tracts had been responsible for drawing Fludd into print, so too a lesser problem — cure by the weapon-salve — was to lead to van Helmont's first publication. The belief that treatment of the weapon causing an injury would heal the wounded person was widespread in the seventeenth century and was based upon the concept of universal sympathetic action in nature.[45] In 1608 Rudolph Goclenius had written a short work explaining that this cure was due entirely to natural causes. The Jesuit, Johannes Roberti, replied in a work ascribing the cure instead to diabolical influence. In the ensuing debate — also widespread and one in which Robert Fludd was to become engaged — Roberti was eventually to write van Helmont for his opinion. The latter rapidly penned his *De Magnetica Vulnerum Curatione* (1621) in which he took to task the views of both Goclenius and Roberti. The former, he wrote, had grossly oversimplified the problem by confounding Sympathy with magnetism alone, thereby judging the phenomenon to be purely natural. But Roberti was no better since he thought the true author of the cure could be none other than Satan himself. Such a view could not be tolerated since the cure is a natural one finding its origin in heaven. It is little wonder then that "Nature . . . called not Divines for to be her Interpreters: but desired Physitians only for her Sons." Indeed, one might think that it was Galileo rather than van Helmont who admonished Roberti to "let the Divine enquire concerning God, but the Naturalist concerning Nature."[46]

The weapon-salve could be explained through the proper understanding of the harmony of the greater and lesser worlds since "all particular things contain in them a delineation of the whole universe."[47] As for Paracelsus, his works were to be praised and the three principles were to be accepted without reservations. Magic is "the most profound inbred knowledge of things"[48] and its basis remains the same whether used for good or evil. Indeed, when one properly understands sympathetic action in nature he becomes aware that the effect of sacred relics differs little from the magnetical weapon-salve itself.—Perhaps aware of the dangerous ground upon which he now trod, van Helmont closed by protesting that "I am . . . a *Roman* Catholick, whose mind hath been to ponder of nothing which may be contrary to God, and that may be contrary to the Church."[49]

The publication of this tract could hardly have appeared at a less opportune moment. Van Helmont's attack on a prominent Jesuit, his defence of magic and Paracelsus — and this combined with his interpretation of the miraculous power of relics — could hardly go unnoticed. Thus, it was hardly unexpected that the Faculty of Medicine at the Louvain should denounce the work in 1623.[50] The following year van Helmont published a second tract on spa waters in which he attacked the views of a highly influential physician, Henry van Heer.[51] Here he made a personal enemy who was to lead the debate against van Helmont over the coming years. Within a matter of months twenty four propositions were taken from the work on the weapon-salve and used as the basis for a charge of heresy. Due to the nature of the charge, the case was tranferred almost immediately from the Civil courts to that of the Spanish Inquisition and late in 1625 these propositions were condemned for heresy and magic and the tract itself was impounded. While van Helmont protested his innocence new censures were forthcoming from the Louvain, Cologne, and Lyons. In the course of this he was twice forced to acknowledge his errors.

In the midst of these personal troubles the letters to Mersenne were written. It is interesting that these indicate little change in his views over the decade and, indeed, when Mersenne addressed him as an adept, van Helmont replied that he was as yet unworthy of this title.[52] His final years were to be beset with constant harassment. The final judgment of the Ecclesiastical Court charged van Helmont with having departed from the true philosophy to espouse superstition, magic and the diabolical art in its stead. Above all, he had followed Paracelsus and his disciples in preaching his Chemical Philosophy, thus having spread more than "Cimmerian darkness" over all the world.[53] He was arrested, his books and papers seized, and then imprisoned. After again having repudiated his errors, he was released under house arrest until 1636 while the Church proceedings continued for an additional six years. In 1642 he obtained permission to publish a medical work on fevers and shortly before his death in 1644 another medical collection appeared in print.

Van Helmont bequeathed a mass of manuscripts to his son with the last request that these be edited for publication. Franciscus Mercurius turned to his task immediately and the *opera* were published first as the *Ortus medicinae* in 1648. This—along with the medical compilation of 1644—was to appear in a total of twelve editions in five languages by 1707.

The *Ortus medicinae* presents a strong plea for reform. It was necessary to "destroy the whole natural Phylosophy of the Antients, and to make new the Doctrines of the

Schooles of natural Phylosophy."[54] Ancient science and medicine was characterized as "mathematical" and logical and this must be avoided at all costs in favor of a truly observational approach to nature. Nor was the ancient approach toward motion any better.[55] Aristotelian local motion had led to a belief in an unmoveable mover. A Christian definition, van Helmont countered, would permit no such restriction on the Creator. In reality motion was inherent in life — implanted in the initial seed by the Creator. If mathematical abstraction could lead to such a mistaken conclusion it could readily be concluded that the Aristotelian descriptive interpretation of nature

> is a Paganish Doctrine drawn from Science Mathematical, which necessitates the first Mover to a perpetual unmoveablenesse of himself, that without ceasing he may move all things . . .[56] Therefore let the Schooles know, that the Rules of the Mathematicks, or Learning by Demonstration do ill square to Nature. For man doth not measure Nature; but she him.[57]

Clearly the new philosophy proposed by van Helmont was one that would seek to reject any concept of nature interpreted primarily through mathematics.

Throughout van Helmont's work may be noted the close association of nature and religion. Once again we are told to look first at the account of Creation in *Genesis.* This — as in the case of Fludd — ascertains the order of Creation and the true elements. Here fire is not mentioned while earth is seen as a product of water. The famous willow-tree experiment and a number of other cases are advanced to show the fundamental elementary nature of water. As for the Paracelsian principles, they are useful since they are obtained by distillation from so many substances, but they are hardly elementary in nature since it is the heat which produces them.

The key to nature is to be found in fresh observations

—and it is chemistry that offers us our greatest opportunity for truth.

> I praise my bountiful God, who hath called me into the Art of the fire, out of other professions. For truly Chymistry, hath its principles not gotten by discourse, but those which are known by nature, and evident by the fire: and it prepares the understanding to pierce the secrets of nature, and causeth a further searching out in nature, than all other sciences being put together: and it pierceth even unto the utmost depths of real truth: Because it sends or lets in the Operator unto the first roots of those things, with a pointing out the operations of nature, and powers of Art: together also, with the ripening of seminal virtues. For the thrice glorious Highest, is also to be praised, who hath freely given this knowledge unto little ones.[58]

Coupled with this was an awareness that quantification — understood here as laboratory weights and measurements rather than mathematical abstraction — might well offer new insight. The willow tree experiment is the best example of this, but van Helmont was also interested in the determination of the specific gravity of metals[59] and the comparison of the weights of equal volumes of urine as a guide to illness.[60] He sought more accuracy in the determination of a scale of temperatures and was led by his studies to insist on the indestructibility of matter and the permanence of weight in chemical change. Of special interest was his reference to the candle enclosed over water. Explained earlier by Fludd in mystical terms, van Helmont insisted that this phenomenon could only be explained in terms of a vacuum in nature.[61]

A thorough-going vitalist, van Helmont proceeded to develop an explanation of the existence of all things based upon his system of elements and their own life cycles. Here he discussed life sources and semina which were responsible for results as diverse as the minerals and human disease. Essential for his explanations were an armory of new definitions. Here the *archeus,* the *ferment* and the *seed* were defined first in terms of inorganic and then organic life. In short, all things arise from water and seed. They proceed to develop from the action of the internal archeus and resultant ferments. And, their life cycles played out, all substances return in time to their primal state, water.[62]

Van Helmont's medicine reflects his background. Unwilling to accept the ancient medical texts, he was also disturbed by those who would accept everything ascribed to Paracelsus. Thus, in these later works van Helmont rejected a doctrine of the microcosm which postulated man as an exact replica of the greater world. Still, this did not prevent him from calling attention to numerous similarities that were to be found in both man and nature as a whole. As an example, he considered disease in man to be a localized phenomenon, and although considerably more complex, similar in many ways to the growth of metals in the earth. Nor was van Helmont less concerned than Fludd with the vital spirit in nature. But if Fludd had sought to isolate this spirit through an alchemical experiment on wheat, van Helmont turned to the distillation of blood — an approach that was to become widespread among late seventeenth century iatrochemists.[63] His belief in the existence of the life spirit in the blood was to be an influential factor in his firm rejection of blood letting[64] while his chemical investigation of digestion was to lead to the concept of acid-base neutralization which he described with both physiological and inorganic examples.[65]

Although many other examples from the *Ortus medicinae* might be advanced to indicate van Helmont's interests, we may forego this to emphasize that although Fludd and van Helmont were both deeply influenced by their Hermetic and Paracelsian background that there were real differences between them. The first sought a new approach to nature stressing inner truths, true religion and a mystical alchemy — the second would not have materially disagreed on these points but was to place a much more noticeable emphasis on new observations. And, regardless of the fact that they shared many beliefs, Fludd and van Helmont were viewed as very different by most seventeenth century scholars. Although near contemporaries in age, their works became known to different generations. Fludd's battles occurred during the last twenty years of his life. The influence of van Helmont was largely posthumous due to the restraint placed upon him by the Inquisition.

For many in the middle decades of the century van Helmont seemed to present a plan for a new philosophy fully as promising as that of the mechanical philosophers. Here was a "Christian" observational approach to nature that seemed to reject the mysticism of the earlier Paracelsians, but still indicated the validity of comparisons made between man and nature. In England van Helmont's work was influential in the many appeals for university reform made during the Interregnum. The possibility of economic prosperity promised by the Chemical Philosophy encouraged others to apply Paracelsian-Helmontian theory to agricultural—and economic—reform.[66]

Still, it remains easiest to follow the influence of the Chemical Philosophers in scientific and medical texts. We find that the earliest publications of the English corpuscularian, Walter Charleton, were pro-Helmontian in tone—even to the extent of including a translation of the controversial weapon-salve tract.[67] John Mayow's much lauded description of nitro-aerial particles proves on closer examination to be little removed from the earlier identification of the vital spirit with an aerial niter.[68] As for Robert Boyle, we find him repeating arguments against the Paracelsian principles in *The Sceptical Chymist* that had been offered by a host of authors from Erastus to van Helmont. His earliest works show a close acquaintance with van Helmont, but even in later texts Boyle indicated his respect for the experimental observations to be found in the writings of the

Flemish physician.[69] As for late seventeenth century medical theory, the work of Thomas Willis — and to a lesser extent Franciscus Sylvius de la Boë — indicate the debt that the chemical physiology owed to the Chemical Philosophy in general — and to van Helmont in particular.[70]

To be sure, the period was one of synthesis — perhaps as many syntheses as there were authors. Yet, as far as a "primal" chemical influence is concerned, the major figure of the period was certainly van Helmont. It would be possible to hint at this through reference to a number of other authors, but it might be of more interest to conclude by reference to one. In a recent paper in *Isis* I pointed to the fact that Isaac Newton had read van Helmont in detail and that there is manuscript evidence to indicate that he had considered his views on the origin of motion and its relationship to the life force. It may be possible to go beyond this. The origin of Newton's Third Law of Motion has always been a subject of interest. Most recently J. L. Russell has referred to the statement by Thomas White in 1657 that whenever one body acts on another, the second reacts with an equal force upon the first. But we are certain that Newton had read van Helmont — and van Helmont had essentially stated — and rejected — the same principle.

The relevant text on motion is to be found in van Helmont's "*Ignota Actio Regiminis*" first published in the *Ortus medicinae* in 1648. It appears in a medical context related to the theory of cure. Van Helmont's rejection of cure by contraries was essentially Paracelsian and a direct consequence of his denial of the humors. The Galenic system had been based upon a "warfare" of the elements in our bodies and, as a result, cure by contraries arose as a necessity. However, van Helmont explained, it was a "mathematical" error of the ancients to assume that "every excess might be reduced to a mean."[71] Those who subscribed to this doctrine accepted a perpetual strife and conflict in nature when instead they should have emphasized its harmony and integrity.

So convinced was van Helmont of the error of the Galenic position that he carried his argument beyond the realm of medical practice to that of the physics of motion. The natural philosophers of the schools affirm "that every patient or sufferer doth likewise of necessity re-act, and for that cause likewise every agent or acter doth re-suffer; neither also that it is any other way weakened."[72] But it is incorrect to charge that any action results in an equal reaction. Indeed, "after I with-drew contraries out of nature, I could not afterwards, in sound judgment, find out any re-acting in the patient, as neither could I admit of hostilities in nature, elsewhere than among soulified or living creatures."[73] In a lengthy section van Helmont offered numerous examples designed to show that under no circumstances is there any relation of action to reaction in the case of collisions. His conclusion: "there is never in these, any re-acting of the patient, or re-suffering of the Agent."[74] His most detailed case is that of the man who strikes an anvil with his fist "and thereby receives a wound." Here the fist is hurt by itself "and therefore neither is there any re-action, as neither action of the Anvil."[75]

Here van Helmont's account must be placed in its medical context — and it suffers from the fact that the correct statement is rejected as Galenic. And, although we do have Newton's summary of van Helmont's other discussions of motion, this manuscript is incomplete and does not include his notes on the *"Ignota Actio Regiminis."*[76] Nevertheless, it seems unlikely that if Isaac Newton took it upon himself to read and take notes on the work of van Helmont that he would have missed this discussion — one that was so closely related to his own interests.

Conclusion

The debates centered on the work of Robert Fludd and Jean Baptiste van Helmont indicate the wide interest aroused by the Chemical Philosophy in the seventeenth century. Fludd's confrontations with Kepler, Mersenne and Gassendi began with his earliest publications (1616, 1617) and continued for the next twenty years. Van Helmont's problems also began with his first publication (1621), but due to official persecution relatively little was known of his work until the posthumous publication of his collected works in 1648. Thus, the European scholarly community was faced with a new, more observationally oriented, Chemical Philosophy at a time when it was just beginning to assimilate the work of Descartes and Bacon. Van Helmont's pleas for educational reform, his rejection of ancient philosophy, and his many observations were clearly noted by a wide spectrum of European scholars at that time.

Reference to the Chemical Philosophers occurs frequently where it might at first be unexpected. However, this is almost inevitable since the Chemical Philosophers did not think of their work simply in terms of chemistry or medicine. Rather, this was a professed attempt to found a "*Philosophia Nova*" that would account for the entire cosmos. Only if approached in this fashion will it be possible to explain the threat to natural philosophy seen in Fludd's mathematics by Mersenne and Kepler, the anatomical question noted by Gassendi — or expect to find anything here that could conceivably be of interest to the Isaac Newton with whom we are most familiar.

We might close by asking whether this material is important for our understanding of the Scientific Revolution. In a recent defence of a more traditional internalist history of science Mary Hesse has argued against the study of Hermetic influences and then gone on to state that "throwing more light on a picture may distort what has already been seen."[77] Surely the study of the large Paracelsian-Helmontian-iatrochemical literature of the seventeenth century will add immeasurably to the complexity of our study of the rise of modern science. Nevertheless, this literature exists — and there is ample evidence to indicate that it was read and discussed. I believe that for this reason alone it must become integrated into our accounts of the Scientific Revolution. A more difficult task — but fully as important — is the determination of the extent to which later "more respectable" concepts may be traced back into this earlier context. Some have been already and others will be, but the fundamental question here is to decide to what extent we are dealing with a simple clash of world systems — and to what extent we have a case of one system influencing the development of the other. This question, a crucial one, cannot be given a fully convincing answer until far more research has been completed.

The Chemical Philosopher as imitator and interpreter of natural phenomena. From the frontispiece to J. B. van Helmont, *Opera Omnia* (Frankfurt: J. J. Erythropilus, 1682).

Frontispiece of Tommaso Cornelio, *Progymnasmata Physica* (Naples, 1663)

Alchemy in the Seventeenth Century:
The European and Italian Scene

CESARE VASOLI

Professor Debus' essay in this volume highlights a number of problems that historians of science have far too often brushed aside with a gesture of annoyance. The study of the hermetic elements at the origins of the scientific revolution leaves many people with a feeling of discomfort, an intellectual malaise, as though something indecent had been discovered and was embarrassingly harped upon. It is, of course, unpleasant to have one's clear account of the growth of science challenged by research into activities long deemed disreputable because they were considered the expression of philosophical "errors" so felicitously superseded by modern scientific "truths". But the clear-cut division between rational certainties and irrational phantasies not only concealed from view important aspects of intellectual life in the seventeenth century, it was blind to the close ties between the general stance of the period and the rise of modern science. Naturally, a historiography that emphasizes only the continuous succession of "true" theories will look askance at any attempt to understand the "non-modern" elements present in the writings of the pioneers of the scientific revolution. One can conjecture that (in an ideal world) science would have progressed more rapidly if it had not been compelled to guard itself at every turn from the fascination of magic. But history cannot exclude the facts, and the facts tell a different story once they are no longer simplified in order to make them appear more intelligible to the contemporary reader.

Professor Debus has shown that the vast Paracelsian and Helmontian literature was read and discussed by the very people who acclaimed the new mechanical philosophy as the dawn of a new age. In the doctrine of Paracelsus and van Helmont, seventeenth century thinkers admired above all the radical break with the dogmatism of the Peripatetics and the Galenists. We must not forget that the "chemical philosophy", especially in the writings of Fludd and van Helmont, is presented as a necessary condition of the general transformation of all knowledge, as the key to a new wisdom destined to enhance all human learning. This ideal, common to most of the representative figures of the new science and the new philosophy, constitutes the general background and the natural support of the scientific revolution. We may smile at the utopia of the Rosicrucians and we can query the accuracy of recent claims about their importance, but we cannot deny the influence of the hermetic tradition on the rise of modern science.

Professor Debus has very clearly underlined the important fact that the "chemical philosophy" was not only a body of scientific doctrines but a general philosophy of nature and of man's place and role in the world. Paracelsus, Fludd and van Helmont appealed to empirical observations and to experiments but the evidence they use was filtered through a web of religious and philosophical ideas that made much of the necessity of discovering the true cipher to the two books of divine wisdom, the Bible and Nature. Such an emphasis derived from the reinterpretation of the hermetic and cabalist writings by the Renaissance philosophers Marsilio Ficino and Giovanni Pico della Mirandola. Pagel may have exaggerated Paracelsus' dependence on Ficino but he has conclusively shown that many characteristic features of Paracelsus' philosophy can be traced back to Ficino's *De vita coelitus comparanda*. The image of the "doctor-magician" in Paracelsus' writings seems largely inspired by Ficino's account of "honest magic". But there is an important difference, underlined by Professor Debus, for Paracelsus appeals to "experience" as the key to decipher the book of nature.

Naturally, Paracelsus' "experience" or "experiment" (the word is the same in Latin) is closer to the hermetic notion than to the one we are familiar with today. It is, in fact, largely based on the macrocosm-microcosm analogy. Nevertheless, there is a real and novel experimental component in the "chemical philosophy". This is one of the reasons why it was so well received and why its condemnation of the empty and impious philosophy and medicine of the ancients did not ring hollow. Wherever educational and institutional reform was a lively issue (be it in Catholic or Protestant countries), the "chemical philosophy" rapidly found adepts and practitioners among those who scorned the psittacine teaching of the Schools.

This broad appeal of the "chemical philosophy" explains the violence of the controversy that raged around it. Not only scientists, but philosophers and theologians joined the fray. One of the foremost opponents of Paracaelsus was no less a figure than Thomas Erastus, the prominent Protestant divine who saw in Aristotelian philosophy a bulwark against the tide of superstition and magic that seemed to bespeak the coming of the Antichrist. Erastus was bent on purifying religion from diabolic contaminations and he vigorously attacked every form of "irrational" belief. The controversy over the Comets of 1577 and 1582 saw him on the side of the champions of freedom against the blind determinism of the astrologers. In his attempt to unseat magic, he denounced both the "nefarious" Paracelsus and his master, Marsilio Ficino, guilty of having given credence to the "disgusting" and "diabolical" myths of the hermeticists, so clearly opposed to Aristotelian reason and Christian faith. In many respects, Erastus' attack merely echoes the thunder of Calvin a few decades earlier.

If Erastus spoke scathingly of the "chemical philosophy", Giordano Bruno described the writings of Paracelsus and Palingenius Stellatus in the most glowing terms. He saw them as the discoverers of the ancient mysteries and as men who had freed themselves from the bondage of orthodox religion. In assessing the arguments proferred for or against the "chemical philosophy" we must keep in mind that in the seventeenth century men still laid more store by theological and philosophical reasoning than at any time since.

Kepler took Fludd to task for failing to see that mathematics is the only source of real enlightenment and that theories can be submitted to quantitative experiment. But once we have recognized the profound methodological differences between Kepler and Fludd we must acknowledge, with Professor Debus, that hermetic beliefs are still at work in Kepler's scientific outlook. His appeal to the musical harmony of the celestial spheres, the roles he assigns to the "soul of the world" and the "living" nature of heavenly bodies are not so far removed from considerations entertained by Fludd. Kepler was intent on integrating hermeticism within his Pythagorean scheme. He wanted to free these ancient truths from their magical shell and show their proper place in the "divine" mathematical vision of the world.

Mersenne's criticism of Fludd is equally revealing. He denounces Fludd's false claims for alchemy but he is mainly concerned with the religious menace presented by this allegedly miraculous science. His bitterest attack is to be found in his theological treatise, the *Quaestiones in Genesim,* published in 1623 at the height of a religious craze when Paris was under the incubus of an "atheistic plot" and menaced by a fanciful invasion of the Rosicrucians. The denunciation of Fludd soon spread to his sources and was connected with the condemnation of Francesco Giorgio Veneto, another major exponent of hermetic and cabalistic ideas. It is clear from this and from analogous incidents that the "chemical philosophy" was often considered a threat to established dogmas. This explains why Gassendi was hostile to it although many of its tenets coincided with those of his own philosophy, and why he distrusted Harvey's method even when he could distinguish it from Fludd's "mystical anatomy".

While engaged in this debate against Fludd, Mersenne became acquainted with the work of Causabon, the Huguenot jurist, who exposed the fallacy of the belief in an ancient hermetic wisdom. Causabon sapped the foundation of the magical tradition but he did not, and could not, damage the "chemical philosophy" in its insistence on experiment and its use of the new atomistic conception of matter. It may even be argued that it was the destruction of the magical overtones of the chemical philosophy that allowed the development of its experimental and practical side.

Professor Debus has established that van Helmont was not operating in quite the same intellectual matrice as Fludd. Of course, for van Helmont, as well as for Paracelsus and Fludd, magic was the most genuine and profound knowledge available to man. He shared their belief in the intimate connection between religious illumination and knowledge of nature, their strong repudiation of Aristotelian and Galenic doctrines and their celebration of the divine character of the "book of nature". But Professor Debus is undoubtedly right when he stresses van Helmont's insistance on experiments, his appeals for a reform of education, his refusal of the magical mysticism of the early Paracelsians, and his vindication of a Christian philosophy of nature against pagan science and medicine. These ideas, novel in so many respects, explain his posthumous fame in a cultural context that was no longer that of Fludd and Mersenne.

The works of van Helmont found their way into the hands of Boyle and Newton and they were one of the means whereby the corpuscular doctrine was disseminated. Thomas Willis and Franciscus Sylvius de la Böe both made much of van Helmont's ideas, and I believe that an investigation of van Helmont's readership would be most revealing. I shall simply say a few words here on the influence of the "chemical philosophy" in Italy, more specifically in the Kingdom of Naples. I leave aside the debt that Campanella, Della Porta, Stelliola and Cesare Coppola owe to Paracelsus to concentrate on the latter half of the seventeenth century.

First of all, I must mention Marco Aurelio Severino's *Antiperipatias, hoc est adversus Aristoteleos de respiratione piscium diatriba* of 1659, a work that is in sympathy with much of Paracelsian medicine. It was attacked ten years later by Tommaso Cornelio for subscribing to the "ridiculous" medicine of the chemical school. Nevertheless in his *Progymnasta* (1687) Cornelio devotes much space to the Paracelsian notion of *"spiritus"* which he discusses along with the Cartesian concept of an active principle of motion. However fond of Galileo and Descartes, Cornelio experienced no intellectual qualms in accepting views derived from Paracelsus.

Giovanni Battista Capucci, one of the leading Neapolitan physicians, praised the freedom of mind of Paracelsus and van Helmont whom he considered more daring than Malpighi. An equally famous surgeon, Lionardo da Capua, praises Paracelsus for his rediscovery of "the ancient way of philosophizing" in his *Lezioni intorno alla natura delle mofete* (1683). Although often described as a Cartesian, da Capua refers to van Helmont as "the great Hermes of Lower Germany" and speaks of him as one of the great "moderns" along with Willis and Campanella. This appreciation of the "chemical philosophy" is found elsewhere in Italy. Magnenus in his *Democritus reviviscens* published in Pavia in 1646 exalts "the defensors of philosophical liberty" such as Tycho Brahe, Kepler, Galileo, Paracelsus and Fludd. In Naples alone, we find a spate of books that promote the mechanical philosophy and yet come under the influence of iatrochemistry and spagiric medicine. The physician Sebastiano Bartoli in a series of writings, which appeared after 1666, displays a thorough acquaintance with the doctrine of van Helmont from whom he borrows the notion that the

auri, et magni lapidis negant plerique, presertim Kircherus et Bet-
tius, asserentes primis illis temporibus fuisse nondum fundendi metalla
habitam, et agendi ea, quae ad Metallurgicam pertinent; qua de
re in Aurifactoria sermo erit: Verum pro certo habendum, fuisse antiquitus
hanc Artem accomodatam ad usum medicum, siquidem liquores stilla-
titios fecisse legitur primi Arabes, et Persae: sicque aliquod recepisse
incrementum. Paracelsus inde Chymiam medicinae penitus associavit, et
peculiarem Paracelsistarum genuit factionem. Crevit demum artis progressus, et Recentiores Medici eandem reddiderunt nobiliorem, rejiciendo
vana omnia, et solum experientiae sistendo, admittentes ea, quae in
pellendis morbis videntur apta: unde plura scripserunt Cartesius, Digbeus,
Borellus, Oliva Sabuco mulier, Thomas Campanella, Robertus Boyle,
Thomas Willis, Franciscus Sylvius de le Boe, Glysonius, Helvetius, Tho-
mas Cornelius, Jo: Bapt. Capucius, Sebastianus Bartolus, Lucas
Tozzi, Carolus Musitanus, aliique non pauci: et inter Anatomicos
Gaspar Asellius, Gulielmus Harveius, Jo: Horne, Jo: Pecquet, Thomas
Whartonus, Ludovicus Bilsi, Natanael Igmorus, Franciscus Glissonius,
Thomas Bartolini, Regnerus Graaf, Carolus Dracassati, Richardus
Lower, Laurentius Bellini, Marcellus Malpighius, etc:

De Chymicae Utilitate.
Cap. 5.

Considerato Chymicae progressu, ejusdem etymologia, et utilitas est
consideranda: et primum dicitur Chymia, vel Chemia, vel Chymatria
à Chamo Egypti Rege, qui salis philosophici creditur inventor: aut
à Chymi verbo Graeco, quod latine est fundo: aut à Ponchymo, nempe succo, vel sapore, quem Chymici è quolibet corpore purum extrahunt.
Definitur autem à Paracelso, quod sit Ars corpora naturalia mixta solvendi, et soluta coagulandi ad medicamenta gratiora,
salubriora, et tutiora concinnanda. Admittit hanc eandem definitionem
Beguinus in Tyrocin. Chym. cujus enim munus est purum ab impuro se-
cernere vel solvendo, vel coagulando, et quibus operationibus perfectio-
ra educuntur medicamenta: spirituosior nempe, et exquisita pars è
corporibus separatur, et qua tota illorum corporum vita oriebatur, quam
tamen partem crassior substantia ex densis particulis compacta con-
servabat; nimirum per hujus crassioris substantiae particulas diffun-
ditur, et sub latibulis veluti intrusa cohibetur, ut explicat Musitanus

Folio 203 from Giacinto Gimma's *Nova Encyclopaedia*, *Biblioteca Nazionale di Bari.*

204

nam hac una Dianę lampade Medicus plus cernit, quàm vulgares = Medici aperto Sole: et Paracelsus dixit eum, qui caret Chymica, se habere ad verum Medicum, sicuti coquus Porcorū ad coquum Principū. Certè quidem si Galenus suo quo Chymicam habuisset, Medicinam dedisset nobis auctiorem, quam fecit, ut ait Jo: Petrus Faber, Mirotech. spagir. et Hippocrates ipse in lib. de veteri Medic. asserit Medicinā fuisse partim inventam, partim inveniendam.

Neque Chymici sūt ex eo improbandi, quod metallicis, alijsq. venenatam habentibus naturam utuntur: etenim Galenici ijdem ea exhibebant absque ulla preparatione, ut constat apud Galenū, Dioscoridem, in Antidotario Nicolai Myrepsici, alijsque locis Autorum, qui crudū mercurium, antimonium, sulphur, et similia in usus posuerūt, quę tamen omia hac arte emendantur, permutantur, intenduntur, remittunturque simpliciū vires, ut inquit Jo: Bapt. Porta.

Verū tamen est huiusmodi medicamēta nisi à perito viro preparētur, damnum potius, quam utilitatē afferre: quoniam Multa sunt, quę nonnisi magis perfectę industrię, et experientię requirūt, nec possūt à tyrone confici, ut inquit Libavius in epist. ad lector. Veteres ideò ne ab ignaris occuparetur ars, quibusdam notis usi sūt.

De Ergalia, seù de Chymicis Instrumentis. Cap. 6.

Ad Ergaliam Chymicorū instrumētorū explicationem pertinere diximus supra, quę cū in magnū excreverint numerū à pluribus Auctoribus diversimodè describuntur, presertim à Libavio, à Mysitano, alijsque artis scriptoribus: Nos tamen nūc notiora subiungimus.

Precipua instrumenta sunt Fornaces destillatorię, reverberatorię et similes, lateribus, et terra pingui, sive figulina, atque arena fimo equino, aut aliqua flamētosa materia, et quandoque salsa aqua composita.

Vasa vel sunt vitrea, vel terrea vitreata, ut aiunt, et quandoque metallica pro destillatione vegetabilium, quę corrosiva carent aciditate. Vitrea autem vasa varias habent formas.

Circulatorium est illud vas vitreū, ubi infusus liquor nūc ascendens, nūc descendens circulatur.

Pellicanus est aliud vas ita dictum à Pellicani figura: = habet à sumitate exitum, per quem liquor infunditur, qui exitus =

Folio 204 from Giacinto Gimma's *Nova Encyclopaedia*, *Biblioteca Nazionale di Bari.*

radiating centre of the *"lux vitalis"* is in the *"precordi"* whence it spreads *"per membranas et partes"*. A disciple of Borelli, Lucantonio Porzio, makes van Helmont one of the main figures of his dialogue *Erisistratus, sive de sanguine missione* of 1682. Giuseppe Donzelli in his *Teatro Farmaceutico* (published in 1713) repeats the Paracelsian doctrine of the three principles (salt, sulphur and mercury) and follows van Helmont in naming water as the material principle on which the "spirits" and the "ferments" operate. Outside the Kingdom of Naples, we find in Costantino di Grimaldi's *Discussioni istoriche, teologiche e filosofiche* published in Lucca in 1721 a praise of such "moderns" as van Helmont, Boyle, Gassendi, Galileo and Descartes. Elia Astorini in his unpublished *Ars Magna* of 1694 mentions Paracelsus and van Helmont as the founders of the experimental chemistry that has produced such remarkable results in Germany and England.

Let me end this brief survey by mentioning two works that I consider representative of the age. The first is Giuseppe Valletta's influential *Discorso filosofico in materia d'Inquisizione* of 1697. Even at this late date Fludd and van Helmont are given honorable mention among the "moderns" who put an end to Aristotelian "servitude", reaffirmed the rights of reason, and restored dignity to scientific research. The second work is a large unpublished manuscript conserved in the Biblioteca Nazionale in Bari. This *Nova Encyclopaedia,* as it is described on the title-page, was begun by the Abbot Giacinto Gimma in 1693 and incorporates hermetic, cabalistic and Lullian elements along with the discoveries of Galileo, Descartes and Boyle. Obviously inspired by the encyclopaedic tomes of Heinrich Alsted, it gives an up-dated version of the "natural magician". The section on medicine draws heavily from the *Encyclopaedia medica* of the Paracelsian physician Johannes Dolaeus. Gimma sees Paracelsus as the founder of the new school of philosophy that has finally united the practice of "true" chemistry with the art of medicine, surely one of the major breakthroughs of the new science. For chemistry, according to Gimma, was only perfected recently by investigators who "relied exclusively on experiments". To all purposes and intents Gimma identifies the "new science" with the "*philosophia atomistica et chymica*". It is the culmination of a progressive development in which Paracelsus and van Helmont played leading roles. The "*ars medicamentorum magna*" is the prototype of Christian learning, not to be set below the achievements of the mechanical philosophy. For Gimma, as well as for many of his late seventeenth century contemporaries, there was still ample room for iatromechanism and the spagiric art even when the hermetic myths on which they had once thrived had been discarded.

New Light on Galileo's Lunar Observations

GUGLIELMO RIGHINI

Galileo's lunar observations extend from 1609 to 1638 when failing eyesight compelled him to abandon his astronomical research. During these three fruitful decades, he discovered or made an important contribution to our understanding of three important aspects of the moon. In chronological order, these are:

(1) the discovery of the mountainous surface of the moon and the first lunar maps;

(2) the discovery of the moon's librations:

(3) the interpretation of the moon's secondary light.

The last two periods are fully documented in Galileo's published writings and in his correspondence, but information concerning the first one is scarce and usually vague. But because Galileo's drawings of the moon are best appreciated in the light of what can be known of the moon's libration, I shall discuss his drawings of the moon in the last section of this paper in which I hope to offer a reassessment of Galileo as a practicing astronomer.

I. *The Libration of the Moon*

Roughly speaking, we may say that the same hemisphere of the moon is always turned towards us. But although this is, in the main, correct, there are small variations at the edge which alter the amount of surface that is visible to us. The moon's equator is slightly inclined (1°32′) to the plane of the ecliptic. Owing to this fact, and the inclination of the plane of the lunar orbit to that of the ecliptic (5°09′), the poles of the moon lean alternatively to and from the earth. When the North pole leans towards the earth we see somewhat more of the region surrounding it, and somewhat less when it leans the other way. This is the *libration in latitude,* and the extent of this displacement is 6°41′. Furthermore, in order that the same hemisphere should be continually turned towards us, it would be necessary not only that the period of the moon's rotation on its axis be precisely equal to the revolution in its orbit, but that the angular velocity in its orbit should, in every part of its course, be exactly equal to its angular velocity on its axis. This, however, is not the case, for the angular velocity in its orbit is subject to a slight variation which is a consequence of the excentricity of the moon's orbit. In consequence of this a little more of its eastern or western edge is seen at one time than another. This second phenomenon is the *libration in longitude,* and the extent of its displacement is 7°57′. Besides, on account of the diurnal rotation of the earth, we view the moon under somewhat different circumstances at its rising and its setting, but the variation amounts to only 57′2″.6. There also exists a true and intrinsic oscillation of the moon which is called the "physical libration" to distinguish it from the former "optical librations". This physical libration is very small, however, and could not have been perceived by the means at Galileo's disposal.

The striking difference between drawings no. 2 and no. 3 in the *Sidereus Nuncius* (see Fig. 1) shows that Galileo recorded, albeit unwittingly, the phenomenon of libration in latitude. But it is only some twenty years later in his *Dialogue Concerning the Two Chief World Systems* that he addressed himself to this problem.

> If the moon did have a natural agreement and correspondence with the earth, facing it with some very definite part, then the straight line which joins their centers would always have to pass through the same point on the surface of the moon, so that anyone looking from the center of the earth would always see the same lunar disc bounded by exactly the

> same circumference. But for anyone located on the earth's surface, the rays passing from his eyes to the center of the moon's globe would not pass through that very point on its surface through which passes the line drawn from the center of the earth to the center of the moon, unless the latter were directly overhead. Hence when the moon is to the east or west, the point of incidence of the visual rays is above that of the line connecting the centers, and therefore some part of the edge of the moon's hemisphere is revealed, and a similar section hidden on the under side; I mean "revealed" and "hidden" with respect to that hemisphere which would be seen from the exact center of the earth. And since that part of the moon's circumference which is on top at rising is underneath at setting, the difference in appearance of these upper and lower parts ought to be noticeable enough because of various spots or markings on those parts being first revealed and then hidden.[1]

It is clear from this passage that Galileo understood the phenomenon of *diurnal libration,* but his remark to the effect that "anyone looking from the center of the earth would always see the same lunar disc" indicates that he had not grasped the consequences of libration in longitude and latitude. Here is how he expresses himself:

> Now the telescope has made it certain that this conclusion is in fact verified. For there are two special markings on the moon, one of which is seen to the northwest when the moon is on the meridian, and the other almost diametrically opposite. The former is visible even without a telescope, but not the latter. The one toward the northwest is a small oval spot separated from three larger ones. The opposite one is smaller, and likewise stands apart from larger marks in a sufficiently clear field. In both of these the variation mentioned already is quite clearly observed; they are seen opposite to one another, now close to the edge of the lunar disc and now farther away. The difference is such that the distance between the northwesterly spot and the edge of the disc is at one time more than twice what it is at another. As to the other

> spot, being much closer to the edge of the disc, the change is more than threefold from one time to the other. From which it is obvious that the moon, as if drawn by a magnetic force, faces the earth constantly with one surface and deviates in this regard.[2]

In 1632, therefore, Galileo was still firmly convinced that the moon always shows the same face to the earth and he sought to explain all the differences in his lunar observations by diurnal libration alone. This is unsatisfactory since the variations can only be ascribed to libration in longitude. The northwest "small oval spot" visible to the naked eye is the *Mare Crisium* and the other spot almost diametrically opposite is the great "circle" known as Grimaldi. Now for the *Mare Crisium* the smallest distance from the edge of the moon is to the greatest distance as 1:2.5 while for Grimaldi the proportion is 1:4.4., in the case of the greatest libration in longitude which is 7°. Naturally, if the *Mare Crisium* is seen closer to the eastern edge of the moon, Grimaldi will be observed further from the western edge (the eastern edge of the moon is the one to the west of a terrestrial observer, namely the one that is illuminated when the moon is waxing).

After the publication of the *Dialogue* and the painful incidents that followed, Galileo, under house-arrest at Arcetri, resumed his lunar observations. On 5 November 1637, he informed his friend Fulgenzio Micanzio in Venice: "My new observations of the moon provide conclusive proof that the movement *("conversione")* of the moon . . . has for its centre the centre of the earth". Two days later, in a second letter to Micanzio, he states his discovery: the moon is not always turned towards the earth in the same way, rather it moves its face in three ways:

> namely it moves it slightly now to the right and now to the left, it raises and lowers it, and finally it inclines it now toward the right and now toward the left shoulder. All these varia-

> tions can be seen on the face of the moon, and what I say is manifest and obvious to the senses from the great and ancient spots that are on the surface of the moon. Furthermore add a second marvel: these three different variations have three different periods, for the first changes from day to day and so has its diurnal period, the second changes from month to month and has a monthly period, and the third has an annual period whereby it completes its cycle. Now what will your Reverence say when you compare these three lunar periods with the three diurnal, monthly and annual periods of the motions of the sea, of which, by unanimous consent, the moon is arbiter and superintendent.[3]

The first variation with its daily period is, of course, the diurnal libration, while the second, with its monthly cycle, is the libration in longitude whose period is the draconic month of $27^d\ 5^h\ 5^m.6$. However, the third variation, to which Galileo ascribes a period of one year, has nothing to do with optical librations. What Galileo is probably referring to is the shift in the position of the line dividing the light and the dark part of the moon which depends on the varying position of the sun and the moon during the earth's annual orbit. It is interesting that Galileo should not have thought of *libration in latitude* and the reason is most likely to be found in his conviction that the path of the moon was perfectly circular. Libration in latitude is a consequence of the excentricity of the moon's path and would only occur to an astronomer who did not share Galileo's prepossessions.

Galileo's important discovery destroyed the age-old belief that the moon always presented the same aspect to the earth. Yet there could still be misunderstandings as when Antonio Santini wrote to Galileo on 3 February 1638 to request more information about "the moving lunar-spots".[4] On 20 February 1638, Galileo wrote a long letter to Alfonso Antonini in which he repeated what he had already communicated to Micanzio but without insisting on the third variation. It would seem that by this time

Galileo had begun to suspect that this phenomenon was different from the other two that he had noticed. Unfortunately, by the end of 1637 Galileo had lost his eyesight and was compelled to cease his astronomical activity.

II. *The Moon's Secondary Light*

In the *Sidereus Nuncius,* the work in which he presented his first telescopic discoveries, Galileo states that he had many years prior reached the conclusion that the moon's secondary light is caused by the reflection of earthlight on the moon. The new device, merely confirmed him in his belief, for the secondary light "shines so brightly that with the aid of a good telescope we may distinguish large spots".[5] The earth and the moon mutually illuminate each other:

> The moon receives more or less light according as it faces a greater or smaller portion of the illuminated hemisphere of the earth. And between the two globes a relation is maintained such that whenever the earth is brightly lighted by the moon, the moon is least lighted by the earth, and vice versa.[6]

Twenty years later this was not yet clear to everyone and in the *Dialogue,* Simplicio rehearses the standard objections to Galileo's account.

> I do not believe that the moon is entirely without light, like the earth. On the contrary, that brightness which is observed on the balance of its disc outside of the thin horns lighted by the sun I take to be its own natural light; not a reflection from the earth, which is incapable of reflecting the sun's rays by reason of its extreme roughness and darkness.

Salviati, Galileo's spokesman, has no difficulty in showing that the moon cannot shine by any intrinsic light since its surface is completely dark when it eclipses the sun. There is one passage, however, that reveals that Galileo had made frequent observations of the moon between 1610 and 1632:

I conclude therefore that just as the surface of the ocean seen from the moon would appear level (except for islands and rocks), so it would appear less bright than that of the land, which is uneven and mountainous. And if it were not that I do not wish to seem too eager, as they say, I should tell you of having observed the secondary light of the moon (which I say is a reflection from the terrestrial globe) to be appreciably brighter two or three days before conjunction than after. That is, when we see it before dawn in the east it is brighter than when we see it in the evening after the setting of the sun in the west. The reason for this difference is that the terrestrial hemisphere opposite to the moon when it is in the east has fewer seas and more land, containing all Asia. But when the moon is in the west, it faces great seas — the whole Atlantic clear to America — a very plausible argument for the surface of the water showing itself less brilliantly than that of the land.[8]

Galileo returned to the argument eight years later when at the request of Prince Leopold de Medici he replied to Fortunio Liceti's *Litheosphorus*. In chapter 56 of this huge Aristotelian tome Liceti sought to account for the moon's secondary light on the analogy of a phosphorescent stone discovered near Bologna. Galileo used this opportunity to state in clear and unambiguous language the reason why the moon was similar to the earth thus ending a debate that had been protracted for four decades.

III. *Drawings of the Moon*

Galileo has been widely criticized for the poor quality of his drawings of the moon in his *Sidereus Nuncius* of 1610. "Galileo was not much of an artist", writes Classen, "and his sketches of the lunar surface bear little resemblance to nature . . . Except for an occasional sea, hardly any feature is recognizable here".[9] Zdenek Kopal, who reproduces two of Galileo's five drawings in his book *The Moon,* is even more critical:

> A mere glance ... will convince us that Galileo was not a great astronomical observer: or else that the excitement of so many telescopic discoveries made by him at that time temporarily blurred his skill or critical sense; for none of the features recorded on this (and other) drawings of the Moon can be safely identified with any known markings of the lunar landscape.[10]

Antonio Favaro, the editor of the national edition of Galileo's *Works,* states that the telescopic observations recorded in the *Sidereus Nuncius* were made between January 2nd and March 2nd 1610. This assertion, for which Favaro adduces no evidence, may be partly responsible for the harsh judgement passed on Galileo's ability to record his visual experience. The visible portion of the moon's disc undergoes periodical variations and the relative aspect of the moon's features vary considerably on account of the optical librations, particularly on days immediately preceeding and following opposition. The date on which an observation is made is therefore of paramount importance in assessing the accuracy of a lunar map.

On August 24th 1609, Galileo presented his telescope to the Venetian Senate and in the accompanying letter to the Doge Leonardo Donati he declared that "it may be of inestimable value for any business or undertaking on land or on sea". Five days later, on August 29th, he wrote to his brother-in-law in Florence that he had shown his telescope to "numerous gentlemen and senators who, though old, have more than once scaled the stairs of the highest campaniles in Venice to observe at sea sails and vessels so far away that, coming under full sail to port, two hours or more were required before they could be seen without my spy-glass".[11]

By the summer of 1609, therefore, Galileo had already perfected his telescope, but there is no indication that he had begun using it to observe celestial bodies. If we are to believe the rumours that Giovanni Bartoli, the Florentine ambassador in Venice transmitted to Belisario Vinta on October 3rd, 1610, Galileo was kept busy making telescopes for the Signoria. This work, even with the assistance of skilled artisans, would have left him little time to gaze at the sky. The first indication that he had begun to make astronomical observations is contained in a letter to a Florentine correspondent to whom he sent his book of *Ephemerides* with the assurance that he did not need it "being for some time already busy at other studies".[12] I surmise that these "studies" are none other than his observations of celestial bodies, but, unfortunately, the extant correspondence is silent on this question until January 7th 1610 when Galileo informed Antonio de Medici that he had observed Jupiter.

In the *Sidereus Nuncius,* Galileo repeats that he observed Jupiter "on the 7th of January" but he does not say when he observed the moon. We know, however, that he made his first observation four or five days after new moon, for he speaks of "the fourth or fifth days after conjunction, when the resplendent moon shows its horns", and it is clear from the context that this was his first observation or at least the first satisfactory one that led him to draw the appearance of the falcated moon. These observations extended over the period of one and possibly more lunations, and since the last lunation of 1609 extended from November 26th to December 26th, it follows that Galileo made his observations during the last months of 1609.

The five drawings in the *Sidereus Nuncius* seem placed in chronological order (see Fig. 1). Drawing no. 1 represents the waxing moon on probably the fourth or fifth day after new moon; drawing no. 2 the moon at first quarter; drawing no. 3 the moon at last quarter, drawing no. 4 the moon a few days before this last phase, and drawing no. 5 the moon once more at last quarter. The diameter of the lunar disc is 80 mm in four cases while it is only 76.5 mm in drawing no. 2, but this difference can be ascribed to the printer rather than to Galileo since such a change bears no relation to variations in the moon's apparent size.

If the moon were a perfectly spherical globe without

Fig. 1. Galileo's drawings 1–5 of the Moon in the *Sidereus Nuncius* (1610).

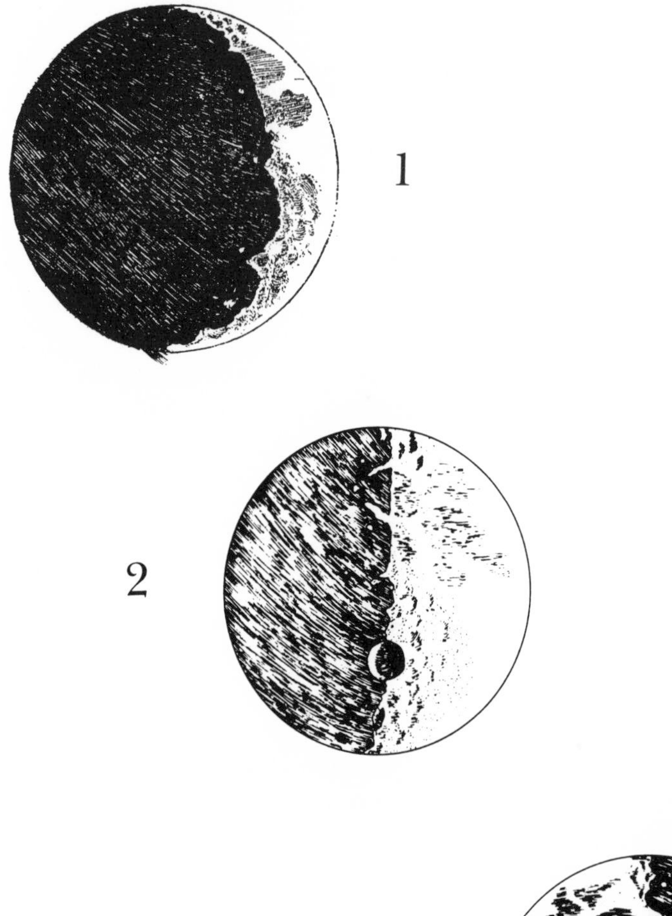

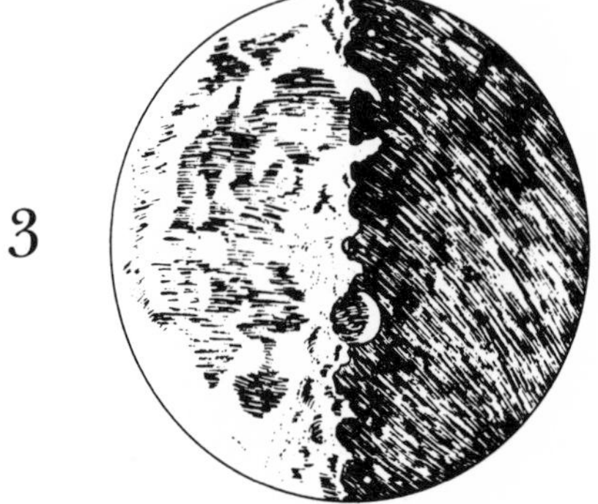

Fig. 2. Diagram of Galileo's first drawing indicating some of the main features of the lunar landscape.

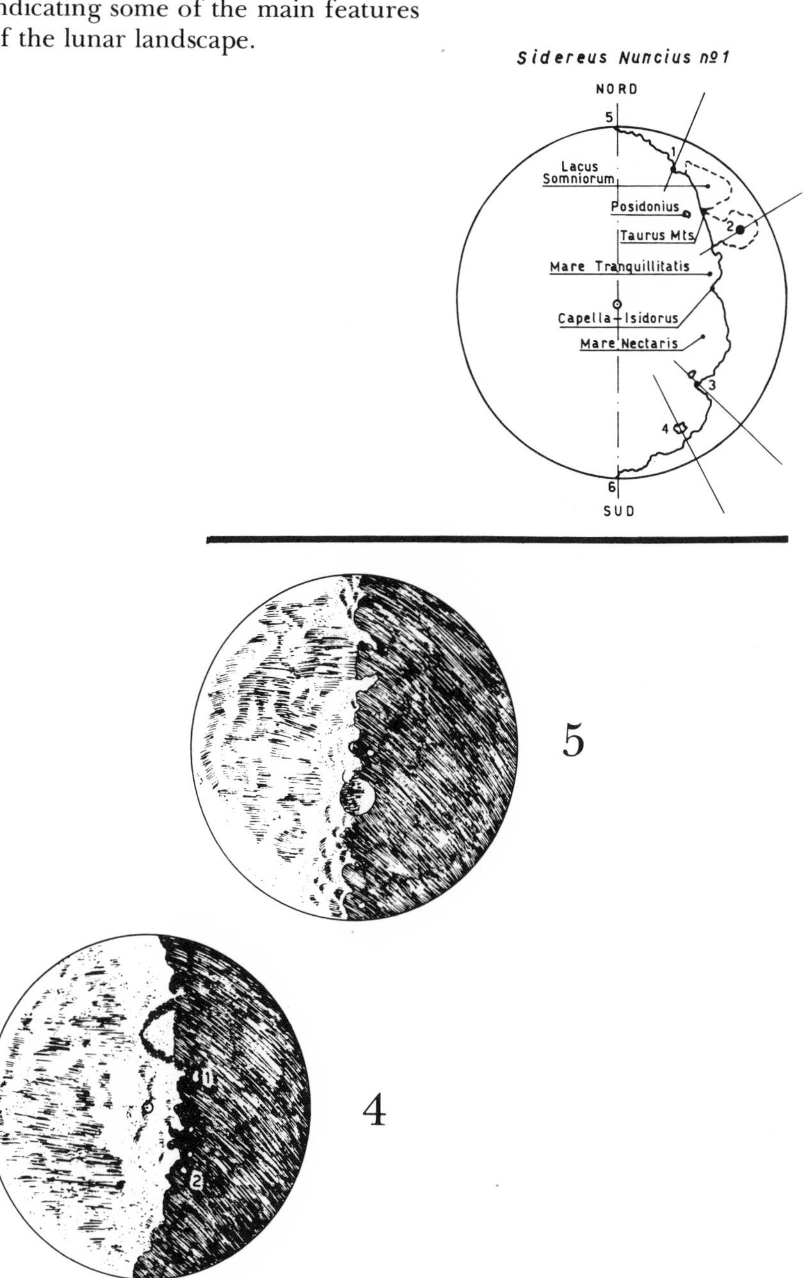

heights and depressions, the age of the moon, namely the period between new moon (the conjunction of the sun and the moon) and the time of a later observation could be determined exactly from the amount of the diameter of the moon that lies in the illuminated part. But since the moon is covered with mountains and craters this is clearly not the case. Furthermore, Galileo had only meager telescopic means at his disposal, and hence the date of his drawings can only be ascertained within a certain margin of error. I calculated that for drawing no. 1, the age of the moon at the time of observation was:

$$4^{d}.62 \pm 0^{d}.08 = 4^{d}\,14^{h}\,53^{m} \pm 1^{h}\,55^{m}$$

I then proceeded to establish the dates when new moon occurred between July and December 1609, and I added in each case $4^{d}\,14^{h}\,53^{m}$, the time interval between new moon and Galileo's observation.[13] This gave, of course, a list of dates on which Galileo could have made his observations. But since the moon is visible after sunset when it is four or five days old, the next step was to calculate the time of sunset and the corresponding age of the moon in Padua where Galileo made his observations. The results are tabulated in the following Table:

TABLE 1

Day, month and time when the age of the moon was $4^{d}.62$; sunset in Padua; age of the moon at sunset.

Dates in 1609 when the age of the moon = $4^{d}.62$		Time of sunset	Age of the moon
5 July	$2^{h}58^{m}$	$19^{h}42^{m}$	$5^{d}.32$
4 August	13 02	19 24	4. 88
2 Sept.	1 17	18 30	5. 34
2 Oct.	16 18	17 48	4. 68
2 Nov.	9 16	17 00	4. 94
1 Dec.	4 10	16 24	5. 13

It is clear from this table that the only day on which the age of the moon at sunset agreed (within a very narrow margin of error) with the age of the moon inferred from Galileo's drawing is 2 October 1609. On that day, the age of the moon was $4^{d}.68$, almost exactly the value of $4^{d}.62$ required by drawing no. 1.

This interesting result is confirmed by calculating the longitude and the latitude (known as the selenographic coordinates) of some of the lunar configurations observed by Galileo. I chose for this purpose six individual points (which are indicated on Fig. 2) and I determined the polar coordinates with respect to the centre of the lunar disc and the vertical diameter that joins the cusps. In Table 2, sin a is the sine of the selenocentric angle which subtends the arc of the great circle that passes through the centre of the disc and the individual point. This is equal to the distance of the point from the centre divided by the radius of the disc. Points 5 and 6 are the two lunar cusps.

The selenographic coordinates of the point O (namely the point where the line joining the centre of the moon and the centre of the earth cuts the lunar surface) are easily determined with the aid of astronomico-chronological tables.

The line of cusps does not usually coincide with the meridian passing through the poles and the centre of the

TABLE 2

Position of 6 characteristic features of drawing no. 1.

Points	Sin a	Angle
1	0.82	23°
2	0.44	60°
3	.68	133°
4	0.79	152°
5	1.00	0°
6	1.00	180°

disc, but its position angle can be computed by means of spherical trigonometry. The data given in Table 2 and the angle of position of the central meridian enable us to determine the selenographic coordinates *l* (longitude) and *b* (latitude) for each point.

The points measured on Galileo's drawing agree quite closely with features on a modern map of the moon. The discrepancies lie well within an acceptable margin of error, and, in the case of point no. 2, identification with the center of the *Mare Crisium* is beyond doubt.

It is not possible to proceed in exactly the same way to date the other drawings. The time that the moon can be observed after sunset at the first quarter (when it sets about midnight) and at the last quarter (when it rises about midnight and crosses the meridian about sunrise) is too large to allow a precise determination of the day on which Galileo made his observations. On the strength of Galileo's statement that he began his observations four or five days after the new moon, it is safe, however, to conclude that drawings 2 to 5 were made after October 2nd 1609, the date of the first drawing. These four drawings correspond to the following phases of the moon: first quarter for drawing no. 2, last or nearly last quarter for drawings nos. 3, 4 and 5. Now between October 1st and December 31st, we have the following choice of dates:
for the first quarter: 5 October, 3 November, 3 December;
for the last quarter: 20 October, 18 November, 18 December.

Let us first examine drawings no. 2 and no. 3 which are characterized by a large circular spot that, in Galileo's words, "lies in the centre of the moon . . . and is much greater than the other ones". This large spot does not correspond to a single lunar configuration but to several large craters in the continental massif of the moon's southern hemisphere. In Galileo's drawings, the diameter of the spot exceeds 350 kilometers and its selenographic latitude is approximately −19°. This corresponds to the area covered

TABLE 3
Selenographic coordinates calculated for drawing no. 1. compared with known lunar configurations.

Individual points	From drawing no. 1		From a modern map of the moon		Lunar configurations
	l	*b*	*l*	*b*	
1	+35°	+40°	+30°	+45°	Group of craters: Plana, Mason, Airy that separate the *Lacus Somnium* from the *Mare Frigoris*.
2	+55	+14	+59	+16	Center of *Mare Crisium.*
3	+27	−37	+23	−35	Continental Massif centered on the crater Rabbi Levi.
4	+19	−53	+17	−50	Groups of craters Baco, Breislak, Clairaut and Cuvier on the southern high plateau.
5	+ 9°	+79	cusp N		The North Pole is beyond the edge and therefore invisible.
6	+14°	−79	cusp S		The South Pole is visible inside the edge of the moon.

by the craters Purbach, Regiomontanus, Werner, Blanchinus, and Lacaille on the southeastern shore of the *Mare Nubium.* The circular structure observed by Galileo is a real feature of the moon and not an optical illusion or the result of bright and shaded areas that can be seen at quadrature and is recorded in drawings 2, 3, and 5, while it is barely outlined in drawing no. 4 which records a slightly earlier phase of the last quarter.

If we compare drawing no. 2 with drawing no. 3 we cannot but be struck by the different distances of this circular spot from the centre of the lunar disc in the two drawings. The arc between the centre of the spot and the centre of the disc is 25°.6 at the first quarter (drawing no. 2), and 15°.9 at the last quarter (drawing no. 3), a variation of 9°.7. On the scale used by Galileo, the difference is 5 mm, an amount that is too great to be ascribed to error or to uncertainty in the determination of the distance. Furthermore, details of the features of the moon surrounding the circular spot all show a "sliding" towards the centre in drawing no. 3 with respect to the same feature in drawing no. 2. It would seem, therefore, that Galileo unwittingly noticed and recorded an effect of *libration in latitude* which he was only many years later to recognize and call the "apparent titubation of the moon". This *libration in latitude* is easily calculated with the aid of appropriate formulae, and for the first and the last quarter between the beginning of October and the end of December 1609 the results shown in Table 4 were obtained.

The difference in libration between drawings no. 2 and no. 3 which, as we have seen, amounts to 9°.7, is accounted for satisfactorily if we assign December 3rd as the date of drawing no. 2 and December 18th for drawing no. 3. The calculated difference of 8°.8 is a fair approximation to the difference of 9°.7 measured on the drawings, and a closer agreement can hardly be expected given the limitations of Galileo's telescope.

TABLE 4

Libration in latitude of the moon at first and last quarter.

Date (1609)	First Quarter	Date (1609)	Last Quarter	First Quarter– Last Quarter
5 October	−1°.7	20 October	−0°.9	−0°.8
3 November	+0°.1	18 November	−3°.4	+3°.5
3 December	+3°.8	18 December	−5°.0	+8°.8

To date drawing no. 4 I used a slightly different procedure. Since many features on this drawing are easily recognizable, I employed their selenographic coordinates to determine the libration of the centre of the lunar disc.

In drawing no. 4, point 1 corresponds to the crater Maurilius and has the following selenographic coordinates:

l (longitude) = +10° b (latitude) = +15°

Point 2 corresponds to crater Apianus and has the coordinates:

l (longitude) = +6° b (latitude) = −28°

The arc (a_1) of the great circle that passes through the crater Maurilius and the centre of the lunar disc is = 20°, and the one (a_2) which passes through Apianus is = 24°.7.

Having the selenocentric coordinates of the two craters and their angular distance from the centre of the moon disc it is possible to compute the libration in latitude. The results of this simple computation suffice to establish that drawing no. 4 was made shortly before the last quarter on 18 December 1609.

Drawing no. 5 is identical in all respects with drawing no. 3. Since during the period we have been examining (1 October – 31 December 1609) there were no two phases of the moon with the same libration in latitude, it seems clear that drawing no. 5 was printed by mistake. It is likely that in his haste to see the *Sidereus Nuncius* through the press, Galileo inadvertently had the same diagram reproduced on two different pages.

In the light of these measurements it is fair to conclude that Galileo was a much better observer than Classen and Kopal have suggested and that he was, in fact, a remarkably faithful recorder of his visual experiences.

Dissertatio cum Professore Righini et Sidereo Nuncio

OWEN GINGERICH

Professor Righini's ingenious paper on Galileo's early telescopic observations of the moon offers me the occasion to reflect more broadly on the relation of observations to theory, especially in the late sixteenth and early seventeenth century. I should like to argue, as many have done before me,[1] that what we see is strongly conditioned by what we have already seen, and (as a corollary) even what we choose to look at is already heavily theory-laden. The image of the astronomer continually surveying the heavens and objectively documenting all he finds as the foundation for future theories is demonstrably false; such a popular view must crumble before any careful scrutiny.

I shall present some evidence showing that the astronomer of the sixteenth or seventeenth centuries was generally far removed from the idealized scientist accruing observations with neutral open-mindedness for the day when some suitable hypothesis would suggest itself. Most observing, then as now, was specifically theory- or goal-directed. Serendipity seemed to play a larger role than usual in the case of Galileo, who could scarcely have realized what lay in store for him as he began his crucial effort to improve the telescope. His discoveries of January 1610 came so quickly that it is difficult to sort out his motivations, which undoubtedly evolved from day to day. He had harbored Copernican sentiments for well over a decade, and after the initial discoveries, this philosophical vantage point cannot be divorced from his choice and interpretation of the further observations. Although Galileo's unique genius prevents us from closely characterizing him by the same pattern as other early scientists, nevertheless I believe that his philosophical presuppositions powerfully shaped his own observational program.

From the sixteenth century there survive innumerable records of wondrous happenings in the sky—eclipses, conjunctions, comets, and the spectacular new star of 1572. On the other hand, *systematic* observations of the stars and planets are so notoriously lacking that in 1665 G. B. Riccoli could aspire to list them all in his *Astronomia reformata.* (Tycho's observations, more than all the rest put together, were not available in print until the following year).

Most astronomers of the early sixteenth century were apparently either satisfied with the status of astronomical theory, or too unimaginative to aspire to improving the state of affairs. A conspicuous exception was Bernard Walther, the associate and successor of Regiomontanus in Nuremberg, who observed regularly with the vaguely delineated hope of reforming astronomy.[2] Beginning in 1471 he established a remarkable series of 746 solar and 615 planetary positions, culminating with an almost daily record of the great triple conjunction of 1503–1504. The internal precision of his measurements was about four minutes of arc.

Copernicus, in contrast, had a specific vision for the reform of astronomy, but a vision certainly not motivated by new observations. He employed the minimum number necessary to establish the orbital parameters, and how many more he actually made is only a guess. In Copernicus' own century his observations, few as they were, were held in high esteem—witness Magini's *Novae coelestium orbium theoricae congruentes cum observationibus N. Copernici* (Venice, 1589), which accepted the observations of *De revolutionibus* while spurning its heliocentric cosmology. A similar attitude was manifested by the Holy Congregation when considering the fate of *De revolutionibus* with respect to the *Index of Prohibited Books.* The argument was made that the book could not be banned outright because its observations were so valuable, and eventually the Inquisition was satisfied with the deletion or emendation of a dozen passages.[3] To a modern astronomer this appears ironic, for he is impressed by the paucity of Copernicus' data rather than its richness. In all, Copernicus reports fewer than 30 new observations.[4] A detailed examination of his 16 new planetary observations reveals the astonishing fact that typical errors exceed half a degree.[5] Had Copernicus adopted Walther's far more accurate observations, his own heliocentric theory might have yielded great progress in practical predictions as well as in the conceptualization of the celestial universe.

The extraordinary nature of Copernicus' discovery was not to wrest from new data a novel conception of the universe, but rather, to rearrange the same old bricks into a new and glorious edifice. As Waldemar Voisé has recently written, "If it is accepted that Copernicus deserves to be called a thinker rather than an astronomer, then it follows logically that he should be called the architect and not the discoverer of the heliocentric theory. To discover suggests a passive attitude toward reality while the term architect signifies a creative attitude, hence one who enriches our knowledge by an active attitude toward the world."[6]

An innovatively different attitude toward observation is evident in the vast undertaking of Tycho Brahe near the end of the sixteenth century. Nevertheless, his observing program was closely coupled with his understanding of planetary theory. Brahe concentrated his observations on special circumstances—near oppositions, elongations, nodal points, apogees, and so on—as may be clearly discerned from numerous notes and comparisons in his observing books. He did not routinely observe each planet's position every night—in fact, he preferred to get some sleep, generally clustering his observations in the evening or before dawn.

A beautiful instance of the consequences of this

1. Lick Observatory photograph of the moon at last quarter.

2. Engraving from *Sidereus nuncius* of the moon at last quarter.

theoretically guided selectivity has been discussed by Victor Thoren,[7] who found that Tycho regularly observed the moon only when full, new, or at the quarters. Thus for many years he missed a very large irregularity in the lunar motion, the so-called variation, until he happened to observe the moon at an octant position in preparation for a forthcoming eclipse. Similarly, when Kepler sought to determine the parallax of Mars by a new method, he could not find the necessary data among Tycho's observations. Concerning this lack of observations, Kepler writes in a vivid passage in his *Astronomia nova,* "Hence I shall use my own, thereby exhibiting to you a ridiculous spectacle . . . I think we put one unit of ten minutes too much, because in order to read the divisions we could illuminate them only with a glowing coal on account of the wind. . . .[Because of the cold] even the bare hands stuck to the iron, so that it was impossible to close the clamps, and with gloves the ruler was

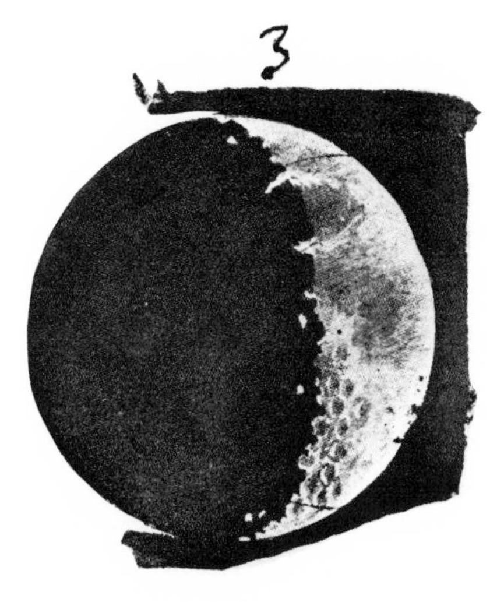

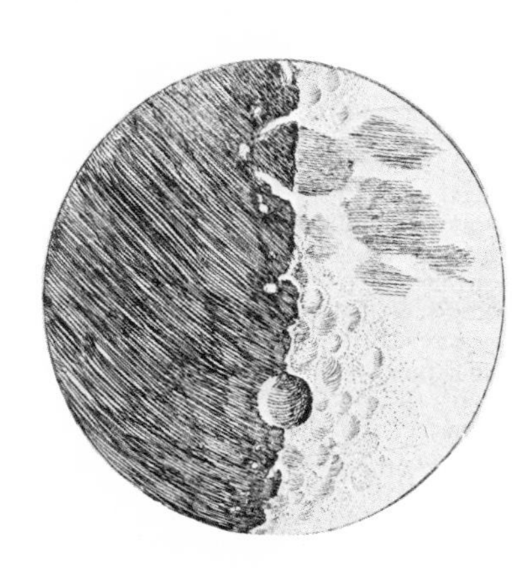

3. Galileo's drawing of the waxing crescent moon.

4. Engraving from *Sidereus nuncius.* Note the similarity of features with the drawing at left.

not securely held long enough for the minutes to be noted."[8]

Clearly, the observations that an astronomer chooses to make will be considerably determined by his theoretical presuppositions and background. Likewise, his interpretation of what he sees after he has chosen to make the observation will be enormously prejudiced by his expectations. In 1604 Kepler predicted and then thought he observed a transit of Mercury. Only after Galileo's observations of sunspots did Kepler realize what he had really seen; he ultimately printed a retraction in his *Ephemerides* for 1617, noting that unwitting he had been the first of his century to observe a sunspot.

The telescopic discoveries of Galileo, on the other hand, appear so stunning and unexpected as to be divorced from the bias of theory. Nevertheless, their interpretation was not instantly unambiguous. The controversy with

Scheiner concerning the nature of sunspots amply demonstrates this point.[9] Even more revealing is the remarkable case of Saturn. When Galileo first observed Saturn through his telescope in July 1610, it appeared in a tripartite form. Late in 1612, after having neglected Saturn for several months, Galileo was astonished to see the planet as a solitary globe. Evidently he supposed that the triple-bodied appearance had been caused by some sort of satellite system, for he predicted its reappearance. By the 1650's a baffling variety of drawings had been published, but no satisfactory model had been proposed. With the blurred historical perspective of three centuries, we generally suppose that the mystery was solved when Christiaan Huygens concentrated a more powerful telescope on the enigmatic planet. But Huygens' solution was done in mind's eye, not at the end of the telescope, for at the time of his discovery, the rings of Saturn were edge-on! Huygens announced his bold conjecture in an anagram, and he waited until the rings began to open again before unscrambling the letters.[10]

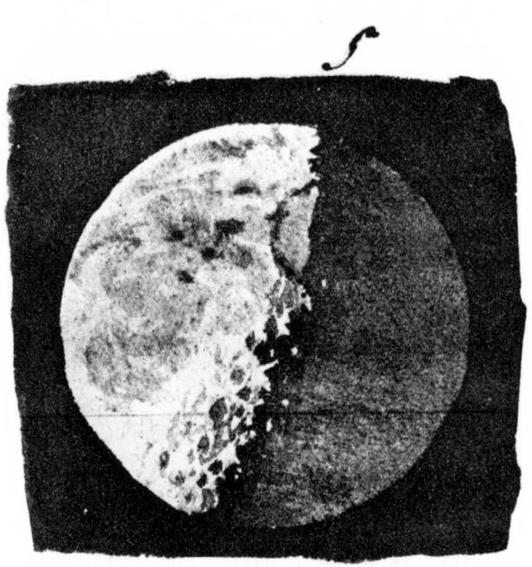

5. Galileo's drawing of the gibbous moon just before last quarter.

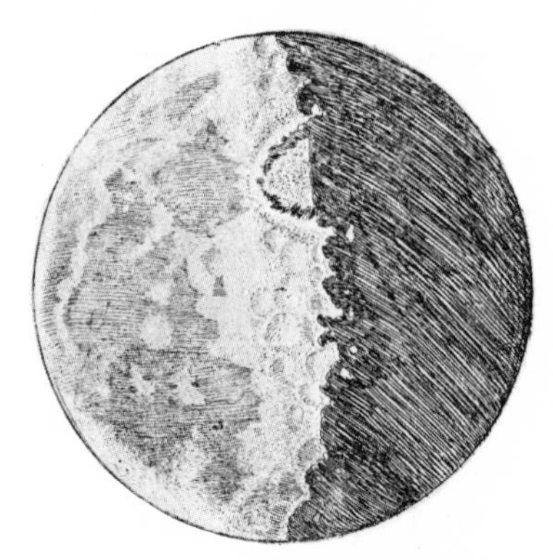

6. Engraving from *Sidereus nuncius*. Note the different placement of the bright points at the terminator compared to the drawing at left.

Just as Copernicus can be called the architect rather than the discoverer of the heliocentric system, we could call Huygens the inventer rather than the discoverer of the rings of Saturn. At the very least, we can say that he invented the model for interpreting the appearance of Saturn—a model so convincing we accept it as reality. And in the same fashion, we could justifiably call Galileo the inventer rather than the discoverer of the Jovian satellites.

On January 8, 1610, the second night when Galileo turned his telescope to Jupiter, he failed to record the conspicuous but distant fourth satellite. Since he had counted three the previous night, his expectations were fulfilled after he had detected the three points of light nearest the planet. Not until nearly a week of observations did he record all four bright satellites. To hypothesize so quickly that the points of light were satellites, a miniature Copernican system, was a brilliant stroke of genius by Galileo, but probably not nearly so obvious as it appears to us in retrospect.

7. Engraving from *Sidereus nuncius* of the waxing crescent moon.

8. Compare this engraving from *Sidereus nuncius* of the moon at last quarter with no. 6.

Again, this hypothesis was a creation or invention on Galileo's part.

We can, I believe, far more readily grant a decisive role to perception and interpretation in the telescopic discoveries of the planets and sunspots than for the observations of the moon. Accustomed as we are to looking at modern photographs of the crater-scarred surface of the moon, it is immensely difficult to imagine that any creative act is required to recognize for the first time the distinctive circular nature of a lunar crater. Yet, the lunar drawings of the *Sidereus nuncius* exhibit a very striking and suggestive hint: only one sharply defined crater is shown, and outrageously oversized at that. Professor Righini has drawn our attention to this feature, which appears along the terminator conspicuously illuminated. It may be seen on the drawings of both first quarter and last quarter, lighted from opposite directions. There is no question but that the crater as shown is impossibly immense. Its diameter exceeds 12°, whereas the largest craters in this part of the moon rarely reach 5°. Furthermore, the curvature of the moon is such that the opposite edges of Galileo's crater could not possibly be so well illuminated at this extended distance from the terminator unless they were incredibly high. This fact, plus the sharpness of its outline, precludes its identification with a cluster of actual craters.

In the case of the Jovian satellites, we can trace the configurations from Galileo's original observing records,[11] through his handwritten manuscript of the *Sidereus nuncius,*[12] to the final printed edition, and, finally, we can compare them with modern recomputations. This process gives us great respect for Galileo's accuracy and for the integrity of the printed edition. Unfortunately, the extant manuscript material does not permit the same rigorous check for the printed lunar drawings. The handmade drawings reproduced with the *Sidereus nuncius* manuscript in the Galileo *Opere* bear an approximate but not precise correspondence to the printed figures.[13] In particular, this very conspicuous crater on the terminator is not to be found. (We cannot even be sure that the manuscript drawings are in Galileo's hand, and in any event it is hardly likely that these are the original sketches made with his telescope.)

In attempting to provide some satisfactory explanation for the large crater in the printed drawing, I examined the appearance of a pair of modern photographs of the first and last quarter moon as seen from various distances. As I increased my distance from the photographs, the smaller craters dissolved into an unrecognizable blur. Eventually only one crater remained distinctly visible as a circle, the crater Albategnius, which lies along the terminator approximately at the position occupied by Galileo's large crater. Because Galileo's telescope was far from perfect—for example, it could not show any satellite within about three equatorial radii of Jupiter's edge,[14] I propose the following hypothesis: On the moon, he could recognize the stark pattern of light and shadows in the mountain rings surrounding Mare Imbrium and he could distinguish the pocking of the bright areas which he described as having "a great number of dark spots. . . This part of the moon's surface, where it is spotted as the tail of a peacock is sprinkled with azure eyes, resembles those glass vases which have been plunged while still hot into cold water and have thus acquired a crackled and wavy appearance."[15] Initially, however, only one definitely circular crater could be convincingly detected. This hypothesis agrees with what Galileo himself wrote: "There is another thing which I must not omit, for I beheld it not without a certain wonder; this is that almost in the center of the moon there is a cavity larger then all the rest, and perfectly round in shape. I have observed it near both first and last quarters, and have tried to represent it as correctly as possible. . . ."

According to my conjecture, the crater Albategnius must have made an indelible impression on Galileo's mind.

Without a micrometer—indeed, without even a proper telescope mounting—a precise map of the moon would have been impossible. Nevertheless, I believe that he recorded the images commensurate with the psychological impact of his sightings. The drawings in the *Sidereus nuncius* do not show any additional truly distinct craters. This is not true of the manuscript drawings in the *Opere,* which do give the impression of numerous but inaccurately positioned craters. At the same time I should like to point out that although there is no manuscript drawing of the moon at first quarter, the drawing labelled "3" made a few days earlier than the first quarter yields a very close correlation between features on its terminator and the one purporting to be the first quarter in the *Sidereus nuncius.* Furthermore, the dark maria are reasonably represented on the manuscript drawing, but bear only a grotesque approximation to reality in the printed drawing. I would like to suggest that the printed drawing is a highly distorted and derivative version of the manuscript drawing.

In conclusion, much as I admire the ingenuity of the method used by Professor Righini to date Galileo's lunar observations, I must regretfully disagree with its validity. His date of December 1609 for drawings 2 and 3 rests primarily on the position of the large crater that I have discussed above. If, as I have argued, Galileo intended this as a symbolic illustration of a crater and not as an exact map, then it cannot be used for specific measurements. Similarly, the rather ambiguous features measured by Professor Righini on the fourth figure of *Sidereus nuncius* appear in quite different positions on the manuscript drawing, thus shattering any confidence in their reliability for quantitative argument. In light of these discrepancies, I am reluctant to believe that the accuracy required for Professor Righini's method can be extracted from the first drawing. (In any event, the age of the moon versus time of sunset is as well satisfied for 29 January or 29 March 1610 as for 2 October 1609.)

As a second conclusion, I believe that both the text and the drawings of the *Sidereus nuncius* support the hypothesis that a single clearly resolved crater, much more conspicuous than the others, loomed large in Galileo's mind and was placed on the printed drawing in disproportionate size to convey his wonder at this feature. Quite possibly this crater provided the basis for his insight into the nature of the lunar highlands. Perhaps on the basis of this and other even fuzzier images, he originally conjectured the existence of a surface crowded with such circular features. (When he wrote the *Dialogo,* he described the lunar craters in substantially more detail; by that time he must surely have had a far more satisfactory telescopic view.) In my opinion, we can not only say that Galileo discovered the craters on the moon, but that at the time of the *Sidereus nuncius* he had invented them in the sense that from a single crater seen well enough he recognized the additional profusion of circular features far more clearly in mind's eye than with his telescope.

Post Script

Since presenting this commentary at the Capri Conference, I have had the opportunity to examine the lunar drawings in *Mss Galileiani 48* at the Biblioteca Nazionale Centrale in Florence. The leaf in question is a stiff folded piece of drawing paper, completely unlike the paper of the original *Sidereus nuncius* manuscript preceding it, or the January 1610 observations that follow. The drawings are carefully made with ink wash, that is, by diluting a brown ink to various strengths. In order to delineate a bright feature separated from the terminator, it was necessary to brush the ink *around* the light spot. In my opinion these figures are not the original drawings made at the telescope, but are copied over from initial sketches.

Six drawings appear together, while another (labeled "8") is on the other side of the fold. This latter drawing is particularly interesting because a bright star is shown close to the dark limb of the moon. The age of the moon in the drawing is $23^{d} \pm 1^{d}$. I have unsuccessfully sought to identify it with a lunar occultation with a planet or first magnitude star visible in Padua in the period August 1609 through July 1610. Venus and Jupiter never have the right longitude for a 23-day moon during that time, and Saturn has the wrong latitude at the one possible conjunction in longitude. An occultation of Mars must have taken place on 20 March 1610 while the moon was below the horizon in Padua. None of the bright stars (Spica, Regulus, Antares, Aldebaran) were closely approached by a 23-day old moon during this time.

Two horoscopes, whose presence is not depicted in the National Edition, share the leaf with the lunar drawings. Neither is dated, but the dates are readily found with a set of planetary tables. One corresponds to 2 May 1590, and the other, unfinished, must refer to the same day even though no planetary positions are given. It is unlikely that this date has any relevance to the lunar drawings. Thus an examination of the manuscript comes no closer to answering the interesting question as to whether Galileo first used his telescope to observe the moon in 1609 or in 1610.

Terrestrial Interpretations of Lunar Spots

WILLY HARTNER

Guglielmo Righini's paper is both intellectually stimulating and aesthetically pleasing. With the sagacity of a master-detective, scrutinizing minute iconographic details and combining them with scanty historical documentation he has ingeniously devised astronomical means of elucidating the earliest phase of modern selenography. And we all know that this marks the beginning of modern observational astronomy. Some of his results are of great historical interest. Let me mention first and foremost the fact that Galileo, in his first observations, had discovered the latitudinal libration—or rather had registered it unconsciously, as Righini terms it—many years before he expounded it in his *Dialogo sopra i due massimi sistemi del mondo.* I admit that the effect may actually have escaped Galileo's attention, but I venture the alternative explanation that in his anxiousness to have his *Sidereus Nuncius* printed so as to secure for himself the priority of the wealth of his discoveries, he was prevented from pointing out details that seemed of secondary importance.

It is these spots on the moon's surface, identified as mountains and valleys by Galileo in 1609, and used as points of reference by later astronomers from Hevelius in 1647 down to Righini in 1974, that I wish to consider.

Leaving aside mythology, such as the Chinese *Yü-t'u,* the "Jade Hare in the Moon", who assists in compounding the drugs which compose the elixir of life, let us turn to the first day of Galileo's *Dialogo.*

Salviati and Sagredo enumerate features of the moon which are similar to those of the earth: it is spherical; it is dark and opaque because it reflects the light of the sun; it is dense and solid, and has an uneven surface, as evidenced by the prominences and cavities revealed by the telescope, and these prominences resemble our most rugged and steepest mountains; the division into brighter and darker areas is analogous to that of the earth's surface into land and sea. In the ensuing discussion the moon's rotation is considered. If it is carried about by an epicycle, it must have a rotation in the opposite sense in regard to its motion around the earth for otherwise it would not always turn the same half of its surface to us. Then follow two more arguments: the earth and the moon illuminate one another reciprocally, as proved by the moon's secondary light; finally, they eclipse one another. The Aristotelian, Simplicio, objects: The moon, although spherical in shape, differs from the earth because it is polished like a mirror. He admits that the moon is not transparent, as claimed by some, but he adds that it has a light of its own, which explains the "secondary light". The earth, being rough, is incapable of reflecting the light of the sun. On the contrary, the material which the moon consists of (the ether) is harder than anything found on earth and is perfectly impenetrable. The alleged mountains, ridges and valleys are illusions created by dark or light spots under the transparent surface, and comparable to coloured spots found in crystals or amber, which evoke the impression of protuberances on the surface. He disclaims the notion that if the earth could be seen from a distance, the land would appear light and the water of the sea dark. The opposite is true: the smooth and transparent waters will appear bright, and the opaque and rough land, dark.

I skip the remaining three counter-arguments, which are of lesser interest, and pass over to the experimental refutation of Simplicio's reasoning.

A mirror is hung on a wall illuminated by the sun, and it appears much darker than the rough wall surrounding it. The opposite wall, which lies in the shadow, is lightened slightly by the illuminated wall. A brilliant spot is visible only where the reflection of the mirror meets the wall. When the flat mirror is replaced by a convex one the reflection on the opposite wall vanishes. What remains, however, is a brilliant spot—the picture of the sun—on the surface of the mirror, which follows the eye wherever one turns. The conclusion drawn by Salviati is that if the moon were as smooth as a mirror—as Simplicio said—then it would be invisible because the light of the sun reflected by a minute part of the moon's surface would not be perceived by the human eye.

Fifteen-hundred years earlier another symposium (equally fictitious) took place on the same subject. Its minutes have been preserved by Plutarch who, although not a scientist strictly speaking, was a man of catholic learning, familiar with the philosophical currents of his time. In his *De facie in orbe lunae (On the Face on the Moon)* the nature of the moon and of its spots is discussed. It is argued that the figure visible in the moon is not only an optical illusion caused by irradiation, for else a similar and much more striking effect would have to be expected of the sun. Neither is it true that it is a reflection of the great ocean surrounding the inhabited earth because this would imply that there are several oceans divided by continents. One is struck here by the similarity with Simplicio's assertion that only a perfectly smooth and transparent body like water is capable of reflecting light. There follows a discussion of the matter from which the moon is made. Is it condensed ether, with or without an admixture of air, as held by the Peripatetics, or is it just a mixture of two of the terrestrial elements, namely, murky air and smouldering fire, as claimed the Stoics? Is it true—and if so, what does it mean—that the earth occupies the centre of an immense, or rather infinite, universe? Can we accept as indisputable the Aristotelian distinction between the terrestrial elements and the celestial ether endowed with eternal and circular motion? Could there not just as well be more than one centre of gravity? But everything points to the fact that the moon has an earthy nature, in particular, that it is neither transparent nor self-luminous as evidenced by its changing phases and by eclipses. "Consequently let us not think it an offence to suppose that she is earth and that it is for this reason that we get the impression of her face. Just as *our* earth has certain great gulfs, so *that* earth yawns with great depths and clefts which contain water or murky air; the interior of these the light does not plumb or even touch, but it fails and the reflection which it sends back here is discontinuous."

A little more than half-way between Plutarch and Galileo, about the year A.D. 1000, the great mathematician, astronomer and optician Ibn al-Haytham (965-1039), better known in the West under his latinized name Alhazen, wrote a treatise entitled *On the Nature of the Spots Which are Seen on the Surface of the Moon.* Only one manuscript, now in the Library of Alexandria, seems to have survived. Since I possess no photostat copy of it, I have to rely on the German translation published in 1925 by my late teacher, the excellent historian of Arabic science, Carl Schoy. Neither before nor after Ibn al-Haytham do natural philosophers seem to find it worth while to inquire into the nature of our nearest neighbour, whence this study deserves our special interest, in spite of its obscure style (which may have to be blamed on the translator) and its tiresome repetitions. Alhazen notes that some believe that the spots belong to the moon itself, while some hold that they belong to a body between the moon and the earth. Others maintain that they are the reflected images of our seas, mountains, and valleys, others, that the moon's surface consists of a transparent layer behind which rough spots of varying colour become visible through the illumination of the sun. All this, however, is wrong, says Alhazen. If the spots belonged to a body other than the moon, they would have a parallax in consequence of which they would be subject to variation, or disappear completely at certain intervals. In particular, the spots cannot be located on the epicycle or on the sphere surrounding it and carrying it about, as claimed by some. The assumption that the spots should be caused by reflection is excluded by the laws of optics, according to which the angle of incidence is always equal that of reflection. The hypothesis that the dark spots are shadows of mountains is untenable since, if this were the case, the length and shape of the shadows would vary according to the relative position of the moon and the sun. But not even the slightest change of this kind has ever been observed.

After a lengthy discussion of density as a measure of colour and of the absorption and reflection of light by different types of matter and by matter of different density, Ibn al-Haytham concludes that the spots which appear light to our eye have a higher capacity of *absorbing* (not *reflecting*) the light of the sun than the dark ones. The denser the matter, the lesser is its absorptive power. Hence the dark spots appear in those places where the matter is densest, impenetrable and, therefore, incapable of retaining light.

As mentioned before, I have not had access to the Arabic original of Ibn al-Haytham's treatise. Since Carl Schoy's translation contains a series of statements that are at odds with those found in Ibn al-Haytham's treatise *On the Light of the Moon,* to which he refers on several occasions, our first task will be to get hold of the Arabic text. For the treatise *On the Light of the Moon,* we have Matthias Schramm's excellent analysis *(Ibn al-Haythams Weg zur Physik,* Wiesbaden, 1963). There we learn among other things that Ibn al-Haytham regards the moon as self-luminous, but that its proper light is very weak. The moon's illumination observed from earth is due to interference between its own light and that received from the sun.

The i dea that the moon is a polished spherical mirror was discussed therefore throughout the ages. It had its last comeback some eighty years ago (my late friend, the astrophysicist Knut Lundmark told me the delightful story), when August Strindberg, on occasion of a visit to the Observatory of Lund, demonstrated to Count Birger Mörner that the spots on the moon clearly showed the outline of the American continent. When Count Mörner modestly objected that at the time in question America was turned away from the moon and could not possibly be mirrored in its surface, Strindberg angrily gave the classical answer: "If one wishes to advance a new hypothesis, one ought to avoid pedantry".

Marcello Malpighi and the Founding of Anatomical Microscopy

LUIGI BELLONI

Translated by Thomas B. Settle

The contribution of the Galilean school to the development of biology has never found a treatment in our literature reflecting its real importance. In particular, commentators have generally undervalued a distinct Galilean iatromechanics, one different from the Cartesian version. Yet we know from its particularly fertile, practical fruits even into the eighteenth century, as for instance in the work of G. B. Morgagni, that for a time there was an unbroken series of students who passed the Galilean inspiration from one generation to the next.

Even now there are many who still believe that the microscope was invented by Antonj van Leeuwenhoek (1632–1723). With all the admiration and respect that I have for the founder of microbiology, it is clear: 1) that enlargement by optical techniques became scientifically valid with Galileo's publication of the *Sidereus Nuncius* in 1610; 2) that the term microscope, like the term telescope, came out of the *Accademia dei Lincei* which published the first representation of a microscopically enlarged image of a biological specimen in its *Melissographia* in 1625; 3) that during the 1660's, when Leeuwenhoek had not yet begun to build and use microscopes, Marcello Malpighi (1628–1694) had already founded anatomical microscopy and had discovered micro-structures in both the higher animals and insects (*De bombyce,* 1669). Malpighi is easily accessible today in my own translations of some of his important texts (1967) and in the excellent edition and commentary of Howard B. Adelmann (1966).[1]

While it is true that Galileo tended to neglect the living

world in favor of the inorganic one, he did play with the microscope and he took "no little pleasure and entertainment" from the observation of "so many little animals". He mentioned this in 1624 when he presented an *"occhialino"* to Federico Cesi (1585-1630), but at the same time he also showed that he was well aware of the services that the microscope could render to biology: "ultimately to contemplate the wondrous grandeur of nature, how subtly it works and with what unutterable diligence".

The microscope was the ideal instrument for achieving this end, the more so because "there still remain infinite and marvellous devices and stupendous mechanisms to be discovered in the make-up of animals and in particular in the human body" as Benedetto Castelli wrote in 1638. Castelli had studied with Galileo and was the teacher of Giovanni Alfonso Borelli (1608-1679) who in turn was the author of *De Motu Animalium* (1680/81), a breviary of Galilean iatromechanics.

Many of the problems discussed in Borelli's work had been posed by Galileo, for example, the hydrostatic equilibrium of fish and the phenomenon of resonance as applied to hearing mechanisms. But even more illustrative of Galileo's atomistic and mechanical bent in explaining the phenomena of life is a passage from the famous section in the *Saggiatore* (1623) on primary and secondary qualities:

> And some of these bodies are continually resolved into minute particles, of which some are heavier than air and fall, and others are lighter than air and rise. This phenomenon, perhaps, is the basis of two other senses, the particles encountering two parts of our body much more sensitive than the skin which latter does not feel the incursions of materials so subtle, tenuous and yielding. Those particles that fall land on the upper part of the tongue, mix with its moisture and penetrate its surface, and cause the tastes, sweet or unpleasant according to the differences in the way their several shapes touch and according to whether they are few or many,

and slow or fast. The ones that rise enter the nostrils and lodge in the mammillulae which are the instruments of smell; thus likewise we receive the stimulations of pleasure or annoyance, according to whether the shapes are this or that, the motions slow or fast, and whether there are few or many.

The Galilean theory supposed that the sapid particles, mixed in the saliva, penetrate the upper surface of the tongue and touch or stimulate the sense receptors that could be hypothesized to be present. The realization of the hypothesis, however required the sort of anatomical microscopy practiced by Malpighi and described by him in his *De lingua* (1665). Malpighi was able to strip off two upper layers of the surface of the tongue, the horny layer and the reticular or mucous layer (the latter still called by his name), and thus expose the papillae, of which one can distinguish three types. He noticed the pores in the lower part of the epithelium, and he supposed that it was through these pores that the particles dissolved in the saliva could reach and stimulate the papillae.

Malpighi followed this with the discovery of the papillae of the skin, as described in *De externo tactus organo* (1665). The point of departure for this discovery was the pig's foot where the tearing away of the hoof revealed a forest of "oblong and almost pyramidal papillae emerging like swords from their sheaths", the sheaths being the openings in the reticular layer that here is particularly thick. He also investigated the papillae in the muzzles of ungulates, especially bovines, where the reticular layer is penetrated to its full thickness by long and thin papillae whose tips come into contact with the horny layer that in this case is very thin.

But these studies of the tongue and skin did not inaugurate Malpighi's anatomical microscopy. I have mentioned them first, out of their real chronological order, in order to call attention to their Galilean connections. Malpighi had been educated in Peripatetic philosophy at the University of Bologna, but he was converted to the "free and Democritean philosophy" as a result of his close collaboration with

Borelli at Pisa from 1656 to 1659 when he taught theoretical medicine at that University. Borelli's house had become a hot-bed of zootomical research in which was discovered, among other things, the seminiferous tubules of the testicle and the spiral fibers of the heart.

After his return to Bologna in 1659 Malpighi continued his anatomical and vivisectional research, always refining his techniques and also resorting to the microscope for help. Thus it was in 1660 that he made some fundamental discoveries with regard to the lungs which he immediately communicated in two brief letters to Borelli. These he subsequently published in Bologna in 1661 under the title *De pulmonibus.* This, in fact, was the effective birth of anatomical microscopy.

According to the traditional scheme the lungs were "fleshy viscera" of a sanguinous nature and a hot-moist temperament. Under Malpighi's investigations they were revealed as an aggregate of membranous alveoli that were connected to the final branchings of the tracheo-bronchial tree and were surrounded by capillary nets, these latter proving to be the long sought link between the arteries and the veins. These discoveries were fundamental from two points of view: 1) the pulmonary parenchyma was proved to have a defineable structure; and 2) the capillaries were shown to bridge the one remaining gap in the circulatory system, thus concluding the story of the laborious demonstration of the circulation of the blood and assuring the general acceptance of the theory, something that had been denied in the first half of the century.

Already notable in these letters, *De pulmonibus,* was the full mastery that Malpighi demonstrated of the methods of anatomical microscopy, for example:

a) with regard to the microscope itself, using instruments of different powers, as the occasions demanded, and observing with both reflected and transmitted light;
b) with regard to techniques of preparation, using desiccation, "boiling", insufflation of trachial-bronchial tree and

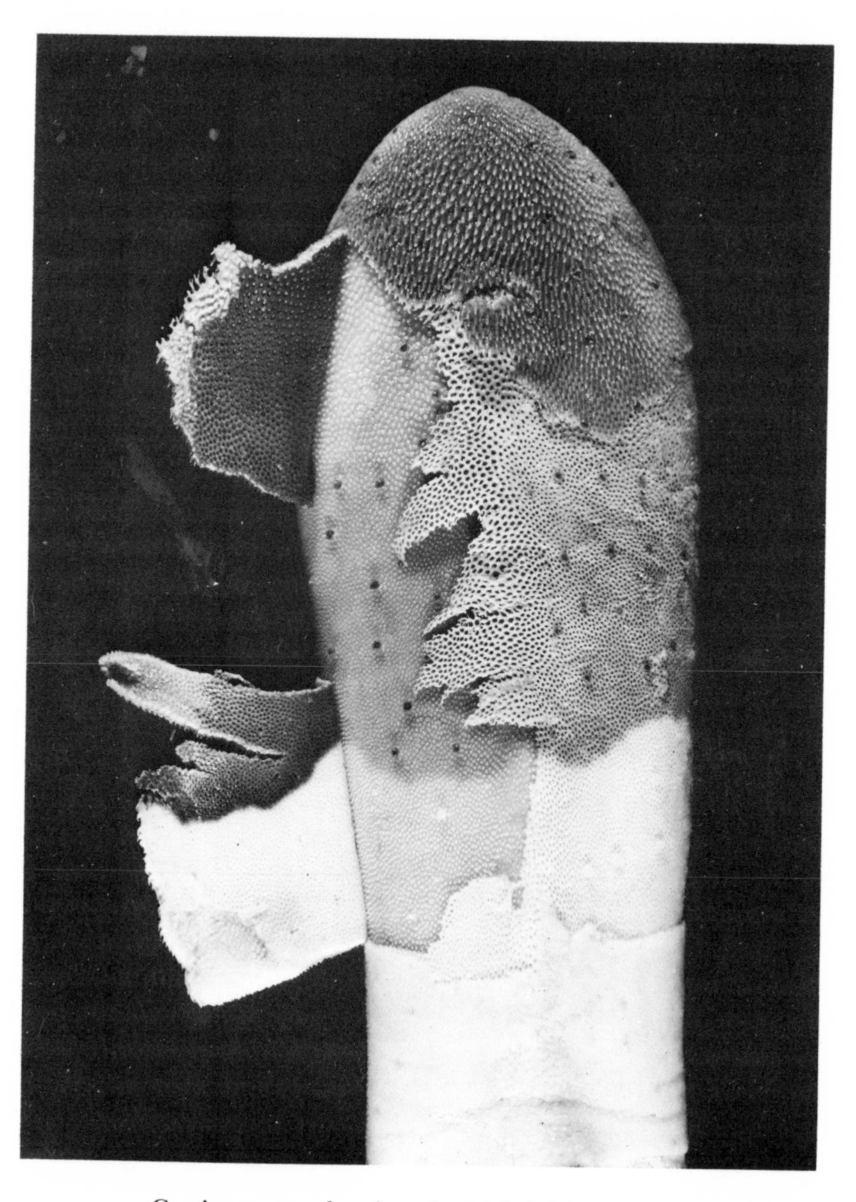

Cow's tongue showing the Malpighian layers

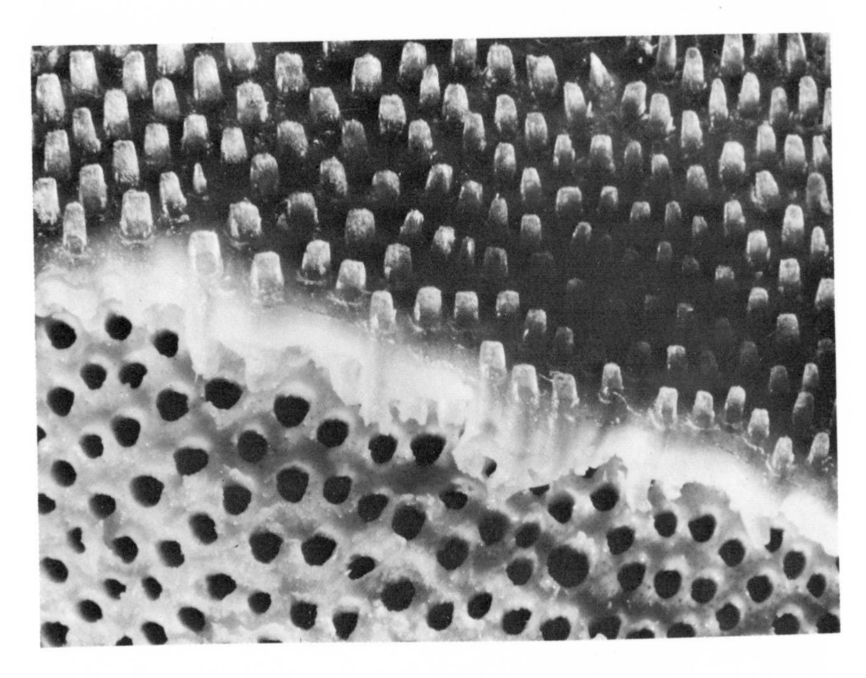

Closeups showing Malpighian layers

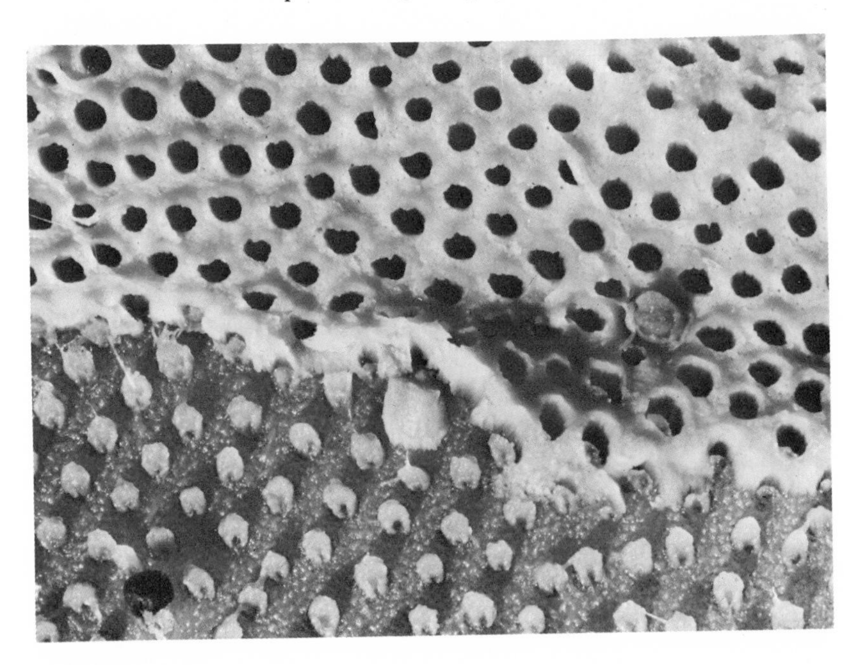

the blood vessels, the draining of blood using vascular perfusion or the expulsion of air by squeezing, corrosion, and so on;

c) with regard to the so-called microscope of nature, using, for example, such an animal as a frog in order to observe with a relatively low-powered instrument a capillary network that in mammals is so fine that it could not be observed with the microscopes then available. The frog is situated somewhat low on the zoological scale. Malpighi very shrewdly observed that nature "usually undertakes its great works only after a series of attempts at a lower level and after testing in imperfect animals its plans for perfect ones." This was an insight that was to have important consequences in the history of biology.

The alveoli, surrounded by a vascular net and undergoing ventilation, were seen by Malpighi as machines well designed to ensure the mixing of the particles of chyle and blood, in other words the transformation of chyle into blood. This process was called "hematosis," and in the Galenic system it was a function of the liver. However, Jean Pecquet had demonstrated in 1647 that the chyle, instead of going to the liver, flowed directly into the blood stream in the superior vena cava, i.e. just before the blood reached the heart for subsequent transmission to the lungs via the pulmonary artery.

In the four years that he stayed at the University of Messina, 1662–1666, Malpighi continued his series of fundamental structural discoveries. To this period belongs the above mentioned research on the tongue and skin; this work in turn started him on a very broad set of neuroanatomical investigations. In the tractatus *De cerebro,* he discussed the white matter of the central nervous system, concluding that it was composed of the same fibers that make up the nerves. He thought of these fibers as very long and narrow tubes, filled with a fluid (the nervous fluid) that was secreted in the grey cortical matter by cortical glands. In his next work, *De cerebri cortice* (1666), he presented his demonstration of the existence of these glands; they are, unfortunately, an artifact of his techniques.

But behind all these findings, true or false, was an attempt by Malpighi to understand the nervous system in terms of a set of "nervous machines" operating from the cerebral cortex to the peripheral terminals. A type of "neuron" would be the basic element of a neurophysiology in which the transmission of a nervous impulse could be regarded as the transmission of a mechanical impulse through a liquid mass according to Pascal's principle.

To this Messinan period also belongs the research that permitted Malpighi to investigate another machine fundamental to iatromechanical atomism: the gland or secretory machine, the mechanism designed to choose certain particles from blood arriving via an incoming artery, separate them from the blood that would then leave through an outgoing vein, and deposit them as a separate fluid in a secretory duct. Just as philology recognizes the same etymology in the terms *cribrum* and *secretio,* iatrochemical theory easily found in the sieve a model for the secretory mechanism, postulating a proportionality of form and dimension between the pores and the particles to be separated. Malpighi recognized that this *minima simplexque meatum structura* was beyond the limits of his capacity to investigate, but he did not give up the search for the structure that could house the pores and that could be localized at the triple conjunction of the ultimate branchings of the arteries, veins and secretory ducts in the glandular parenchyma.

Malpighi located his secretory machine in a follicle that on one side is continuous with the secretory duct and on the other is embraced by arteries, veins and nerve endings. Both the canals that connect the arteries to the veins and the contiguous glandular follicle are permeable to the particles that have to be eliminated, but they are not permeable to the blood particles. This is because they have pores that correspond to the size and shape of the former but not of the latter. The venules, however, do have the size and shape which permit the transmission of veinous blood. Thus, secretion could be explained in purely mechanical terms, without invoking vitalistic faculties. Malpighi presented his best and most convincing findings in favor of this scheme in the *De renibus.* Dyeing by ably performed affusion had revealed to him the renal tubules properly so-called (both straight and contorted); and dyeing by arterial injection had revealed the clumps of small vessels attached to the branches of the interlobular arteries. And these findings suggested to him that the ampullar ends of the tubules were enclosed within the vascular tufts (the Malpighian corpuscles).

Having established the continuity of the circulatory system, suggested the nature of the mixing machine for hematosis, put together various parts of a nervous mechanism furnished with sensitive receptors and constructed a secretory machine, Malpighi undertook to analyse the universal liquid without which all these devises would be for naught. The result was his hematological tractatus *De polypo cordis* appearing as an appendix to *De viscerum structura* (1666).

The polyps of the heart had already been noticed in the bodies especially of persons who had died of gross cardio-respiratory insufficiency, and they had even been included with some frequency in anatomical drawings. While previous authors had variously explained the growth of these polyps using traditional humoral theory, Malpighi saw in them the result of a process of coagulation taking place internally comparable to the coagulation of blood external to an organism. The study of the coagulum was therefore fundamental, and Malpighi finally demonstrated that the "phlogistic crust", in spite of its whitish color, was derived only from blood which latter "with its purplish component confuses our poor eyes". To accomplish this, Malpighi had subjected blood to the same analytic techniques that he had used with such success for viscera and organs; he thus succeeded in separating artificially the coagulum and serum in an already hardened mass, a separation that occurs spontaneously when fresh blood is allowed to stand in a container. After repeated washings the coagulum, "from the intensely red-black that it was, became white, while the water that had extracted the colored particles became red". The phlogistic crust corresponds in substance to the red-black coagulum when the latter had been washed out; the difference between them is only quantitative (in the proportion, that is, between the white and the red) and not qualitative as was required by the humoral theory.

Even a microscopic observation of a piece of the coagulum shows the two components: a) a knot of white fibers that are evidently derived from the coming together of other, even smaller filaments (such as happens in the crystallization of salt); b) a red liquid that impregnates the fibrous mat which is resolved under the microscope into a myriad of red atoms. This discovery of the red globules is Malpighi's own, in spite of the fact that it is otherwise attributed in a general literature that ignores the clear description in the *De polypo cordis.*

So even an abnormality such as the polyp can furnish important clues to the genuine and normal operations of nature. Just as aberrations can be more illuminating than marvelous and refined natural mechanisms, so the "study of insects, fish and the first and unelaborated outlines of animals, in the course of [embryonic] development, has served in our century for the discovery of more than was learned in all previous eras during which investigations were limited to the bodies of perfect animals." This is a methodological consideration and the program that inspired both the *De bombyce* (1669) and the embryological and botanical works published by the Royal Society in London in the 1670's.

The embryological research was inspired by the practice of artisans who "in the construction of machines usually first make the individual parts in order to see separately the pieces that later would be put together." Malpighi says this in *De formatione pulli in ovo* (1673); actually he had already followed in the research described in *De bombyce* the work of artisan nature in each of the three stages (larva, chrysalis and adult) through which it forged the organic machine of the silkworm. This is a machine with distinct components such as the air ducts (tracheae) or the duct with pulsating centers (corcula) for the propulsion of the blood.

Thanks to this embryological tract and to the following *Appendix* (1675), embryology acquired a refined structural content and became an important aid in illuminating adult morphology. Again, just as the study of insects and fish illuminated the morphology of the perfect animal, the study of embryos is so much more fruitful because artisan nature constructs the parts separately before putting them together. In the embryo the miliary glands that come together to form the liver are still distinguishable, like the small sacs that remain distinct and constitute the liver in crustaceans. Both embryogenesis and phylogenesis follow paths that are bound to cross.

For Malpighi the fetus grows in the same way that the small plants contained in seeds grow: from an enveloped state it develops and increases in size as a result of the nutrients liquified by the warmth of the nest and by the fermenting process initiated at fertilization. This development gave fuel to the doctrine of preformation which was then the widely held alternative to the theory of epigenesis.

Malpighi's principal embryological findings were: the vascular area with the terminal sinus, the cardiac tube and the various segments that compose it, the aortic arches, the somites, the plicae and neural tube, the cerebral vesicles, the optic vesicles, the protoliver, the glands of the prestomach, feather follicles, and so on.

The analogical-comparative method followed by Malpighi found its most ample formulation in the introduction to the *Anatomes plantarum idea (A General Anatomy of Plants)*:

> the nature of things, hidden in darkness, is revealed only by the analogical method *(cum solo analogismo pateat)*. For which, one must work in such a way as to be able to *take apart* the mechanism under investigation by having recourse to other mechanisms that are simpler and more easily accessible to sense experience. But the things that are more important and perfect, those that are more necessary for human use and more worthy of consideration, attract human genius first.

Even the interest of Malpighi was aroused in this way; in the fervor of his youth he had immersed himself in anatomical investigations of the more perfect animals.

> But these, hidden in their own shadows, remain in darkness; so it is necessary to study them analogically, using simple animals *(simplicium analogismo egent)*. Thus, I began to investigate insects; but even this had its difficulties. In the end I reverted to the investigation of plants, and after a long exploration of this world I again took the road back to my first studies, starting from the level of vegetable nature. But perhaps not even this will be enough, given that precedence is due to even a more simple world, that of the minerals and elements. At this point the project becomes immense and absolutely out of proportion to my powers.

If Malpighi stopped when faced with the work which would be involved in systematically studying minerals, he nevertheless confronted with extraordinary success the vegetable kingdom. The *Anatome plantarum*, appearing in London in two parts in 1675 and 1679, assured him the title, otherwise divided with Nehemiah Grew (1641-1712), of the founder of the field of microscopic anatomy of plants. For him even plants were structured mechanically: they are traversed by tubes in part analogous to insect tracheae, and they show a basic "cellular" structure, derived from aggregations of utricles, the same as already described with the name *cellulae* in the *Micrographia* (1665) of Robert Hooke (1635-1703). these structures would acquire a very different physiognomy following the elaboration of the cell theory in the nineteenth century.

Malpighi used the notion of the microscope of nature very fruitfully in his investigations of the anomalies and sports of nature, the monstrosities and pathological abnormalities. In the wart, for instance, dermal papillae appear visibly enlarged. And in addition, the anomaly shows the structure in such a way as to make the single elements appear distinct, thereby shedding light on the normal state of affairs. In the ectopic horn in cattle one can see the consolidated elongations of papillae distinctly isolated on the surface, whereas in the normal horn these are hidden because they are fused in the single scales from which the horn itself derives.

In addition, the correspondence between the normal and abnormal horn is morphogenetic; the metamorphosis of the cutaneous layers into the structures of the horn is provoked by a mechanical stimulation that usually is exerted by a bony growth but which in abnormal conditions results from the irritation of the yoke and the subsequent gathering of the exhudate in the subcutaneous tissue. Analogous mechanical morphogenesis is also seen in the polycystic kidney: the glandular follicles (Malpighian corpuscles) appear large and distinct because they have been dilated by urostasis following a blockage in the effluent duct. And again, liver and lymphatic follicles are enlarged, respectively, in the nodules of a cirrhotic liver and those of a lymph gland under the effects of a disease such as tuberculosis.

Malpighi had planned to collect a case book of pathological anatomy, and we can find specific indications of this in his *De polypo cordis.* He had practiced dissections on "the bodies of the dead in order to investigate the causes and the results of various sicknesses".

Under the influence of the microscopical discoveries of Malpighi the importance of local lesions became more recognized in the 1660's. And the discovery of minute mechanisms that together make up the living body gave to the study of abnormal structure more significance than just a simple means of learning about normal structures. The breakdown of these mechanisms—even if studied only at the macroscopic level, as for instance with regard to observable lesions—disturbs the organism and is translated into clinical manifestations proportional to its location and nature.

From the microscopical anatomy grew the investigation into the seats and causes of diseases, to paraphrase the title of the book by G. B. Morgagni, *De sedibus et causis morborum per anatomen indagatis,* 1761. Morgagni was a student of two disciples of Malpighi, Ippolito Francesco Albertini (1662–1738) and Antonio Maria Valsalvia (1666–1723).

Malpighi also contributed notably to the study of plant pathology, in particular to the understanding of plant galls. These appeared to be the product of a disease; actually they are the result of structural changes induced by the egg-laying of certain insects. This work thus completed that begun by Francesco Redi (1626–1697) in 1668 towards the disproof of the theory of spontaneous generation of insects. Redi, of course, was a member of the Galilean school.

Finally, Malpighi was the author of an important methodological work, *De recentiorum medicorum studio,* in which he upheld the notion of a rational medicine against the empiricists. This "rational medicine" was built into his *Consulti,* a testament to the career as a medical practitioner that he carried on alongside his great biological work.

In this quick presentation of Malpighi's work I have drawn attention to the importance of the microscope, a scientific instrument since 1610 by virtue of Galileo, and the micro-structural set of mind and mechanical atomism that matured in the ensuing half century. On this background

Malpighi built his anatomical microscopy, using both analogical comparisons from the realms of insects, plants and (at least potentially) minerals and experimental anatomy, including not only vivisection, but also a whole repertory of other experimental skills.

Malpighi's work was exquisitely experimental but it was exposed to the weaknesses not only of arguments from analogy but also of the possibility of optical and experimental artifacts. Three centuries afterward we can describe his errors so easily that it would be ungenerous if we did not point out some of the varied and sometimes grave traps that lay on a path that, given retrospective clarity, seemingly ought to have been avoided with the greatest of ease and security.

There is the case of the hairs that Malpighi observed in the prisms of dental enamel; these, I expect, are only understandable in terms of an optical illusion caused by the undulations of the prisms themselves. Fortunately Malpighi was seldom fooled by optical illusions; he knew how to avoid the illusory images created in the compound microscope, images that were deadly in the microscope until the beginning of the nineteenth century, but which were already documented by one of Malpighi's contemporaries, Thomas Willis (1621-1675).

The typical artifact, however, was the glandular structure Malpighi found in the cerebral cortex *(De viscerum structura,* 1666). The two techniques that produced this structure, the delamination described above and the tinting by affusion, had given brilliant results in the research on the tongue and skin, on the one hand, on the kidney, on the other. In this case, one takes a fresh brain, boils it gently, and while it is still warm removes the pia mater. One then puts a drop of ink on the bare surface. After a short while one removes the drop and observes the preparation enlarged under reflected light. The cortex does, indeed, appear to be made up of glandular follicles. We now easily recognize these as artifacts.

Malpighi was even trapped by the microscope of nature, the tool he normally used so well: for example in his discovery of the capillaries in the lungs of frogs. In the case of the "tactile hair" in horses and cows, however, he saw it; but he saw it as architypically simple, whereas in fact it is a complex structure.

Morgagni, the continuer of Malpighi's work, knew very well the dangers I have just listed. In fact, he had

> "a praiseworthy fear of the deceptions possible in too powerful microscopes, and of injections and other techniques of preparation. He wished to investigate nature in the open . . . rather than constrained by artifice, even though sometimes he was served by that artifice. He always hated to transfer to humans the implications of anatomical observations made on animals; but he dissected a great number of animals for the sake of comparative anatomy".

Malpighi, Descartes, and the Epistemological Problems of Iatromechanism

FRANÇOIS DUCHESNEAU

Luigi Belloni and Howard B. Adelman have made the outstanding contribution of Marcello Malpighi (1628–1694) familiar to historians of seventeenth century biology and medicine.[1] In this paper I wish to draw attention to some epistemological problems that are imbedded in Malpighi's theories and that are also at the root of Cartesian iatromechanism where, as I hope to make clear, they receive a different treatment. Methodological differences are closely connected with these problems and I do not believe that we can understand, for instance, the change in the concept of physiology from Boerhaave to Haller without examining them.

Belloni contrasts two kinds of iatromechanism: the Cartesian and the Galilean, and he sees an example of the latter in Malpighi's microscopic anatomy. Malpighi's method is characterized by (a) the use of microscopes (with various magnifications and different types of lighting), (b) techniques of anatomical preparation, and (c) "the microscope of nature", i.e. the natural enlargement of structures that can be obtained by selecting particular organs under normal or pathological conditions. For instance, the "microscope of nature" is at work in the observation of the network of capillaries in the frog, of renal cysts, of the morphogenesis and metamorphosis of cutaneous strata in abnormal horns.

Malpighi's method is inspired and guided by two epistemological principles. The first is his "free and Democritean Philosophy" which Professor Belloni rightly specifies as an "atomistico-mechanical conception", grounded in Galileo's celebrated distinction between primary and secondary qualities in *Il Saggiatore* of 1623. Galileo had used this distinction to build a theory of sense-impressions that implied the geometrical structure and the mechanical operation of sense organs. Hence, for an iatromechanist like Malpighi, the idea that anatomic structures conceal minute and complex machines and that the arrangement of these accounts for organic functions. For instance, the membranous alveoli of the pulmonary parenchyma are conceived as a machinery capable of mixing the particles of the chyle with those of the blood. It follows that the haematosis as a function should be deducible from the specific structure of the machinery. Even more significant is the model for glandular mechanism, grounded in the representation of a follicle. On the one hand, the follicle ends up in an excretory canal, on the other hand, its surface is covered by a network of capillaries and nerve ducts. The congruence between the shapes of the pores and those of the particles coming from the apparent vascular system accounts, on the analogy of a sieve, for the phenomenon of secretion. The hypothetical character of this explanation is underlined by Belloni in his introduction to Malpighi's *Opere scelte.* He mentions that Malpighi determines *a priori* the possible connection between the ultimate ramifications of the ducts which get entangled in the glandular parenchyma: arteries, veins, and excretory ducts. Malpighi must renounce the elementary geometrical structure of the glandular meati, because it proves to be inaccessible to observation. However, he presupposes that their composition is truly adapted secretory mechanism, and he undertakes a series of anatomical observations to determine the modalities of the process for various conglobated glands. Although the atomistico-mechanist principle is, in this way, used in the explanation of phenomena, it does not suffice to account for Malpighi's iatromechanism.

The second epistemological principle behind Malpighi's method is a principle of order which governs the analogies throughout the three kingdoms of nature. This principle reduces the manifold of particular observations to an *a priori* explanation. Its meaning is clear from the following passage from the *Risposta del Dottor Marcello Malpighi alla lettera intitolata "De Recentiorum Medicorum Studio dissertatio epistolaris ad amicum":*

> Nature operates according to a necessity which is ever uniform. So the wisdom of man is never so dulled that it cannot succeed in unveiling a large part of its contrivances. Indeed we may look at and admire the discoveries of Astronomy, in relation to meteors, the causes of which the human mind has penetrated. Besides, we progress by framing hypotheses about rainbows, rain, ice, and even lightning, which unfortunately we experience to be more cruel than other things of nature. We can make the same statement about the mechanisms of our bodies, which are composed of strings, thread, beams, levers, cloth, slowing fluids, cisterns, ducts, filters, sieves, and other similar mechanisms. Through studying these parts with the help of Anatomy, Philosophy, and Mechanics, man has discovered their structure and function. Then, resuming the *a priori* process, he has succeeded in constructing models with these. Such models enable him to make the causality of the given effects conspicuous and to give an *a priori* reason for them. With the above and the help of discourse, he apprehends the way nature acts, and he lays the foundation of Physiology, Pathology, and eventually the art of Medicine.[2]

This passage contains, in my opinion, a justification of what Professor Belloni calls *"la mentalità micrologico-strutturistica"*, which is linked with mechanistic atomism. But it affords an even more fundamental indication of the nature of Malpighi's medicine.

Mechanistic models, according to Malpighi, imitate technological devices and so can teach the real causes of phenomena. *Anatome subtilis* discloses the concealed structure of parts and this is sufficient to discover the corresponding function by means of mechanistic analogies. The relation between structure and function is secured and a necessary knowledge of natural processes becomes possible in the case of living beings. This knowledge is then equated with geometrical knowledge because it accounts *a priori* for the causation of vital effects. Let us note here that there is no room for any clear-cut hypothetico-deductive method: Malpighi cannot guard against deceptive analogies and premature generalizations, as for instance, when he assimilates the cortical structure to a network of glands. This is easy to understand since structural and operational analogies underlie his physiological explanations. By reducing the principle of life to local motion Malpighi dealt with *anatome animata*. His grounds for this reductionism were not spurious for mechanical models are useful in showing that nature is homogeneous at the levels of normal observation and of micrological structures. Vital phenomena are no longer opaque and may be treated in the same mathematical way as mechanical phenomena. If there is any difference with mechanics, it lies in the boundless use of what we might call "structural schematism". Vital phenomena exhibit a complex self-regulation and one must go beyond the explanation of organic elements to account for the integration of elementary structures: thus, for instance, "schematism" appeals to a comparison with a clock. In such models, the explanation of the motive power remains defective, and the same can be said of the dynamics of fluids which is presupposed in the account of circulatory mechanisms.

The explanation of living functions also depends on the notion of a natural order to which all mechanical processes are subordinated. This order reveals itself in the unified scheme which prevails among vegetal as well as animal productions. The observations collected in the *Anatome plantarum* (1675–1679) aim at bringing out the mechanical structure of vegetals. Malpighi draws comparisons between the vegetal ducts and the tracheas of insects; he interprets the cell-structure discovered by Hooke as a dispositional state for mechanical effects, like the minute machinery composing the organs of animals. The specificity of vegetal structures and functions, as such, does not hold his attention whereas the unity of the scheme of nature does, and this unity constitutes the principle of constant order which prevails over the plurality of phenomena. Hence, for instance, the following justification of his analogical method:

> Nature, wrapped up in darkness, discloses itself only through the analogical method. Therefore we must scour it all so as to be able to analyze the more complicated machines by means of the simpler and more accessible to sense experience.[3]

The same conception of natural order is to be found in the embryological theories which Malpighi develops in *De formatione pulli in ovo* (1673). When applied to the successive stages in the formation of an embryo, anatomy shows the particular parts out of which nature, like a mechanic or craftsman, contrives the organic whole. This is another occasion for speaking of the microscope of nature. However, the growth of the embryo is a process by which a structure prior to incubation increases quantitatively, but the structure itself is induced in the egg by a fermentative emanation caused by the male seed. The development of the organism is assimilated to a kind of mechanical phenomenon, but the original production is ascribed to the order of nature which is a kind of transcendent mechanism that accounts for analogical constancy among phenomena, and allows for the repudiation of all forms of epigenesis or chance-process in the production of organisms. Jacques Roger has suggested in *Les sciences de la vie dans la pensée française au XVIIIe siècle* that this explains why preexistence and, later on, preformations tie up so easily with a physics based on Democritean ideas. The new methodology required a principle of order, a nature capable of maintaining its structures.

Along the same line, it is interesting to note that such a conception of nature does not only justify the unity of a schema according to which all living beings are made, but also the constancy in causal processes by which all phenomena are mechanically produced. It follows that in Malpighi's theory of living being, the animation of physiological processes is irrelevant. The problem fades into the ineffable, and the new iatromechanistic science has no need to take it into consideration:

> I know that the way whereby the soul makes use of the body to operate is ineffable; however, in the operations of vegetation, sensibility, and motion, the soul is necessarily determined to operate according to the machine to which it is applied. This same way a clock or a mill is equally moved by a lead or stone pendulum, or by an animal, or by a man, and even if an angel moved it, he would produce the same motion by change of place that animals do produce, etc. Therefore, since I do not know the process of operation in an angel, but I do the exact structure of the mill, I would figure out this motion and action, and if the motion of the mill were disturbed, I would try to repair the wheels and their damaged composition, leaving aside the study of how the motive angel proceeds to operate.[4]

The methodology specific to Malpighi's analysis of living processes is based on epistemological principles which explain the merits and shortcomings of Italian iatromechanism from Borelli to Baglivi and, later, of the physiology which Boerhaave developed at the beginning of the eighteenth century.

Different epistemological choices determined the path of Cartesian iatromechanism.

Cartesian biology is generally connected with the theory of the animal-machine and the model of automata. However, *automata* have the status of an "hypothesis", in the Cartesian meaning of this term which is linked with Descartes' notion of phenomena and the method of description. Phenomena are defined as the *"effets qui sont en la nature, que nous apercevons par l'entremise de nos sens"*.[5] As for the method of description, Descartes assumes that one can conceive (or deduce) more effects than those which are actually contained within our world. In the course of a single life one cannot proceed to an intellectual inspection of the entire real world, hence a brief description *(brevem historiam)* of the main phenomena is fully justified in order to know the effects of which we seek the causes. It is a question of linking the notion of true cause with the idea of a constant effect given in sense perception. This is where the hypothesis intervenes, for one has to determine the significant phenomena, those which exemplify most clearly the order and economy of nature, and to devise a model capable of replacing the deduction from the true cause which cannot be reached directly. The last sections of the *Principia* define moral certitude, and compare the discovery of a code which enables us to make sense out of a series of signs with the explanation of all phenomena by means of a deduction from presumed causes.

> But they who observe how many things regarding the magnet, fire, and the fabric of the whole world, are here deduced from a very small number of principles at random and without good grounds, they will yet acknowledge that it could hardly happen that so much would be coherent if they were false.[6]

To solve a cryptogram we must possess a key. In the same way, the hypothesis from which we deduce phenomena must correspond to some formal conditions: it must enable us to account for given particular effects in such a way as to comprehend them all, and it must be inserted within a set of principles that can bestow on it "*une certitude plus que morale*" (*plusquam moraliter*).[7] The certainty of metaphysical principles extends to mathematical principles and to those which afford proper ground for deducing mechanical laws. Hence, the hypothetical character of a deduction consists only in using a history (description) of significant phenomena to direct the deduction towards its end, *i.e.*, the explanation of physical effects by their necessary causes. It follows that reason should finally substitute an indubitable account for the tentative explanation of phenomena by means of hypothetical models. Meaning is given to the elements according to the way they combine, but the principles which actually determine the combination are not known. They are inferred from the appearances. For Descartes this means building a coherent model, capable of symbolizing the unknown order in the cryptogram. If we knew the law of formation for all possible cryptograms (provided that the message significantly represents the combination of the cipher) then we could raise our hypothesis to something more than moral certitude. This notion of "hypothesis" helps to understand the kind of mechanism Descartes uses in his biological theory.

From an epistemological point of view, the Cartesian theory of the animal-machine is well illustrated in the *Traité de l' Homme,* which Descartes wrote around 1632, and which was published posthumously (a Latin version in 1662, the French text in 1664). The following passage is significant:

> I suppose that the body is nothing else than a statue or machine made of clay, which God has framed deliberately and rendered as similar as possible to a human being: this way, He gives it the external colour and shape of all our members, but he also places in it all the parts needed to make it walk, eat, breathe, and thus imitate all our actions which can be thought of as produced materially and depending only on the disposition of organs. We see clocks, artificial fountains, mills, and similar mechanisms, which, though made by men, possess nevertheless the force of moving themselves in a variety of ways. And it seems as if I cannot imagine so many different movements in this mechanism which I take to be made by the hands of God, that I would be unable to imagine that there might be even more.[8]

It is clear from this passage that the problem is to establish a theory of medicine on the basis of the sole laws of motion. Although these parts have a complex structure and a subtle mechanical agency, they are reducible to geometrical intelligibility. If a mechanism can move, it only needs an impulse to start the articulated interplay of its parts. Georges Canguilhem is fully justified in asserting in his book *La formation du concept de réflexe aux XVIIe et XVIIIe siècle:*

> *On ne saurait trop insiter sur le fait que l'assimilation des fonctions physiques à de purs et simples phénomènes mécaniques entraîne Descartes à réduire au contact, au choc, à la poussée et à la traction, toutes les relations que les parties de l'organisme soutiennent entre elles. C'est dans la recontre de cette affirmation de principe et des observations anatomiques dont il croit pouvoir se contenter qu'il faut voir la raison dernière de la conception systématique que Dieu se fait du mouvement animal.*[9]

Descartes states clearly that vital motions are to be reduced to the laws of mechanics. A relation can be seen between the necessity of a geometrical interpretation and the explanation of anatomical structures as mechanisms.[10]

Teleology is excluded from the study of biology since organic functions are deduced from the sole mechanical agency of parts. There can be no appeal to souls or spiritual agents as the efficient causes in the production of the motion of these parts.

Descartes was anxious about his arguments and he tried to sharpen them, as can be seen in the *Description du corps humain* written in 1648, and his letter to Henry More of 5 February 1649. In the *Description du corps humain,* Descartes dismisses the relation between the movements of the body and the action of the soul in the theories of Aristotle and Galen as an anthropomorphic prejudice which stems from the fact that we can control some of our bodily movements. This prejudice is linked to ignorance of the fact that the body's structure is complex enough to imply some sort of auto-regulatory, cinematic agency "for considering nothing but the exterior of the human body, we have not imagined that it had in itself enough organs, or springs, to move itself by itself, in as many ways as we see it move itself".[11]

But there is no real unity between mind and matter. In the end, as is well-known, Descartes has to appeal to the extreme artificiality of a physiological correlation between mental operations and movements of animal spirits at the level of the pineal gland. Descartes was led to deprive animals of any properly reflective activities. This followed from his metaphysical bias but had as well some tenuous basis in experience, namely in the possession of a language learnt without any aptitude for being trained or for expressing any "natural impulses". In spite of such difficulties which were linked with his conception of man, Descartes felt fully justified in developing a theory of the mechanical autonomy of vital functions even if his argument could only be based on the claim that this theory was more intelligible because it dispensed with animal souls. Here is an example of his reasoning:

> Since the other functions that some attribute to it [the soul], as moving the heart and the arteries, digesting meats in the stomach, and suchlike, which do not contain in themselves any thought, are only bodily movements, and that it is more usual for a body to be moved by another body, than it be moved by a soul, we have less reason to attribute them to the latter than the former.[12]

Supplementary reasons are found in convulsive movements that take place without the intervention of the will and indicate that bodies have a mechanical disposition which is sufficient to produce effects analogous to those of voluntary activity. Furthermore, death seems to follow from some mechanical deficiency rather than the departure of the soul from the body. Descartes makes it clear, however, that he is speaking of probable processes:

> It is true that we may have difficulty in believing that the mere disposition of organs is sufficient to produce in us all

> the movements that are not controlled by our thoughts: this is why I will attempt here to prove it, and to explain in such a way the whole machinery of our bodies, that we will have no more reason to think that it is our souls which excite in them the movements that we do not experience to be led by our will, as we would have for judging that there is a soul in a clock, which causes it to tell the time.[13]

The letter to Henry More, dated February 5, 1649 illustrates even more clearly the hypothetical character of his theory about animal-machines, as well its axiomatic value in deciphering organic phenomena. Here again Descartes mentions the prejudice which has persuaded us since childhood that animals think (childhood being the model of a state which prevents or inhibits the construction of a genuine ·science of nature and of "animated" beings). We infer that animals think because we believe in a unitary principle of movement which acts on bodies, but we must distinguish two principles of movement. The one is corporeal *("plane mechanicum et corporeum, quod a sola spirituum vi et membrorum conformatione dependet"),* the other incorporeal and belongs to the thinking substance. Given this distinction, where does the movement of animal spirits come from? Descartes' reply is significant: "I soon saw clearly that they could all originate from the corporeal and mechanical principle, and I thenceforward regarded it as certain and established that we cannot at all prove the presence of a thinking soul in animals".[14] In other words, acts which are said to spring from a spiritual spontaneity in animals can be explained as the effects of some mechanical agency. "It is more probable that worms and flies and caterpillars move mechanically than that they all have immortal souls".[15]

All animal movements can be explained without recourse to thought, merely by the disposition of the organs. Furthermore art imitates nature and man can make various sorts of automata, and "it seems reasonable, since art copies nature, and men can make various automata which move without thought, that nature should produce its own automata, much more splendid than artificial ones".[16] There is no way of assigning a relationship between thought and the structured parts of the body. In animals, the distinguishing aptitude to being trained and natural impulses remains heterogeneous to thought, and depends directly on the structured parts of the body. Just as a true automaton presupposes mechanisms for transmitting or converting movement and mechanisms which guarantee a relatively autonomous source of energy, so life in animals may be equated with heat from the heart.

In so far as animals possess sensory awarenes, this depends on the structure of their bodily organs and on the mechanical changes to which they are susceptible in relation to the phenomenon of sensory awareness. Even if there is no real "reflexology" in Descartes, as G. Canguilhem has shown, nevertheless there is a very advanced theory of the mechanical relation to be established between changes in sense organs and "automatic" movements of response. We can therefore conclude that the mechanistic theory of living beings propounded by Descartes has the status that he assigns to "hypotheses". In establishing this hypothesis, "automata" serve as a model, or rather the different types of automata are so many models that can be used to represent the structure and the functioning of living bodies. In a famous passage in the Fifth Part of the *Discours de la Méthode,* Descartes claims:

> This will not seem strange to those, who, knowing how many different automata or moving machines can be made by the industry of man, without employing in so doing more than a very few parts in comparisons with the great multitude of bones, muscles, nerves, arteries, veins, or other parts that are found in the body of each animal. From this aspect the body is regarded as a machine which having been made by the hands of God, is incomparably better arranged, and possesses in itself movements which are much more admirable, than any of those which can be invented by man.[17]

Descartes' model takes several forms. The living body is compared to automata of different types: clocks, artificial fountains, mills, church organs, etc. . . ., in sum, to systems which differ as much by the internal structure of their parts as by their movements. Moreover, the model of the clock serves more specifically to represent the autonomy and regularity of the functioning of the living body, independently of any principle of inventive spontaneity, and independently of any "universal instrument which (might) serve for every sort of explanation", such as reason. "It is nature which acts in them according to the disposition of their organs, just as a clock, which is only composed of wheels and weights is able to tell the hours and measure the time more correctly than we can do with all our wisdom".[18] Hydraulic or pneumatic models represent the circulatory process of the blood and animal spirits, which assure vital coordination. Thus Descartes compares the neuromotory system to a church organ and its pipes; he compares it also to the hydraulic system of fountains; and the two models are combined in the celebrated passage from the *Traité de l'Homme,* where the circulation of animal spirits in the brain and nerves is interpreted according to the mechanism of musical fountains. Simple mechanical tools, such as cords, levers, wedges, pulleys are used to explain the transmission of sensations and the operations of muscles. The production of motor energy calls for the model of a fire-engine, the heart, whose mechanism is released by a chemical process of fermentation analogous to that of hay or wine. Even from Descartes's point of view, the inadequacy of the models can hardly be denied. What is the point of his comparing the flameless fire in the cardiac cavity with the spring that drives a clock? M.D. Grmek has captured the nature of the Cartesian enterprise when he writes:

> If a human or animal organism is interpreted in a Cartesian way, as an automaton, one cannot escape the logical necessity

> to suppose divine intervention by the First Engineer. A complicated machine must be built by some superior intelligence. Then two possibilities can be invisaged. Let me express these in modern language: an animal-machine is either an automaton with cybernetic regulations and something like a program-tape inserted in it by the First Engineer; or it is a kind of car, or better a very complex factory, which cannot operate without permanent intelligent conduction and supervision. Descartes chose the first logical possibility, which was certainly very audacious on his part, for in his time nothing was known of feedback circuits and program records. We can now easily understand why he was not able to express clearly all the meaning of his beast-machine analogy: he was in search for a still nonexisting mechanical model.[19]

Grmek's modern analogy must be accepted with caution for Descartes presupposed the constancy of the creative act, after the creation of the machine, in the internal processes of the automaton: if the system is self-sufficient at the level of differential motor activity, it is because the engineer continues to make the machine work, by using the same causal power which produced the machine. This aspect seems to have escaped many interpreters of Descartes' theory of animal-machines, perhaps because the doctrine of continuous creation is itself ambivalent. The world is an ordered succession of mechanically related phenomena; the laws of nature do not presuppose any other teleology than the divine "fillip". Yet, at the same time, the succession of phenomena and the mechanical laws which explain them illustrate a creative act, a transcendent teleology, which renews itself at each moment. What holds for the system of the world, holds *a fortiori* for systems which depend on it in the order of deduction, from the simple to the complex. Organisms belong to a more phenomenal world than the system of the heavens but stem from the same final transcendent cause. If the explanation of each function requires the intelligibility of a strictly mechanical chain of cause and effect, the explanation of organic structures is reducible to such a mechanical process. For Descartes, the problem becomes crucial when he has to produce an explanation of animal generation according to natural efficient causes. If we suppose that animal-machines are cybernetic systems with self-regulating devices, self-regulation must intervene both at the level of preservation and at the level of reproduction of the system, for the two are linked in the biological order of things. Whence the importance of embryology in Descartes's eyes, for it seemed to him that his solution would become the touch-stone of his whole physics.[20]

Jacques Roger has described the evolution of Descartes' thinking on this point. In the *Discours de la Méthode* Descartes admitted the failure of his explanation in the *Traité de l'Homme.* Descartes was unable to follow in biology the method he employed in physics and which he formulated as follows:

> If we consider that God is the Allmighty, we must think that everything He did was possessed from the beginning of all the necessary perfection. However, in order to know the nature of plants and men, it is better to consider how they progressively grew from seeds, than to consider how they were created by God at the beginning of the world. In the same way, we will explain the better what is generally the nature of all the things which are in the world, if we can imagine some principles, most intelligible and most simple, from which we will clearly show that the stars, the Earth, and all the visible world could have been produced, as if from some *seeds* (even though we know it has not been produced this way): that would be a better explanation than if we were satisfied with describing the world as it is.[21]

Descartes had already described this method in Part Five of the *Discours,* but it proved useless in the explanation of the bodies of animals and particularly of human bodies.

To deduce effects in an orderly way, causes must be adequately conceived, and only the order of rational investigation can lead to certainty. But an adequate notion of the structure of seeds was beyond Descartes' ken in 1637. Because of this lack of knowledge, Descartes interprets mechanical epigenesis on the model of an "automaton":

> I contented myself with supposing that God formed the body of man altogether like one of ours, in the outward figure of its members as well as in the interior conformation of its organs, without making use of any matter other than that which I had described, and without at the first placing in it a rational soul; excepting that He kindled in the heart one of these fires without light.[22]

But, even from Descartes's point of view, this methodological procedure, which supplements the *a priori* explanation with an *a posteriori* one, is not completely satisfactory. First, the results are no more than hypothetical. Second, since the starting point is life, such as sensation reveals it, the conceptualized structure of the living body is not free of anthropomorphic bias. The organs are interpreted with the help of notions derived from man's relationship to the instruments he creates and restricted to the possibilities inherent in automata. This point of view also defines their functional relations as well as the relations between the entire organism and its milieu. A text from 1648 shows that it is with such *ad hoc* principles that Descartes finally built his theory of generation. The passage is from *La description du corps humain,* and deals with "mathematical embryology":

> If we happened to know well the parts of the seed for each animal species in particular, for instance, for the human species, we could then deduce the shape and structure of each of our members by mathematical and certain reasons; and reciprocally, knowing some of the features of this structure, we could deduce then the structure of the seed.[23]

What commentators have too often neglected in this passage is the equivalence Descartes establishes between the *a posteriori* and the *a priori* approaches, and the fact that inference can be based on the knowledge of many aspects of organic phenomena that result from epigenesis. It seems that Descartes finally arrived at the paradox of submitting biological theory to the rigor of a mechanistic explanation (valid for the entire domain of physical reality) in the very act of describing biological entities as beings of a special type. But are not 20th century biologists ensnared in the same paradox when they interpret genetic metabolisms with the help of cybernetic models? Can such a paradox be avoided? The main deficiency in the Cartesian theory comes from too strict a dependence on "automata" which are deficient with respect to self-regulative power and especially with respect to the processes of growth, decline, conservation and reproduction. Moreover, automata are incapable of accounting for the proper life of the parts. A correct theory of living fibers would ruin the Cartesian model constructed on a central source of motor energy.[24] But within the framework of the Cartesian theory itself, the methodological principle of the animal-machine cannot be sustained without an appeal to teleology.

G. Ganguilhem points out that the Cartesian theory rests on two postulates: "*Le premier, c'est qu'il existe un Dieu fabricateur, et le second c'est que le vivant soit donné comme tel, préalablement à la construction de la machine*".[25] In fact before being applied to the model, the theory requires an infinitely powerful efficient cause to produce that infinitely complex and perfectly regulated automaton which is the animal body. But the theory also requires that the living organism be the preestablished end of the mechanical process both at the level of our explanation and at the level of God's production of organisms according to mechanical means.

One knows that the true cipher of a cryptogram has been found when one reads with its help the meaning of the message. In the same way, the hypothesis that God made automata on the model of living bodies is acceptable if it can be made to work. But the deciphering of a cryptogram presupposes a grid which is based on a general semantic theory. The hypothesis of the animal-machine is based on a model (that of automata) which is directly linked to a certain stage of development in human technology. In other words, the norm of intelligibility which allows the hypothesis to be fashioned is, in this case, a kind of "technological anthropomorphism" which relies on teleology. The attempt was brilliant and audacious, but epistemologically fragile. In particular, the technological models available to Descartes did not assign a sufficient reason to the organic functions, whether it be case of individuation, conservation, or reproduction. These models supply only a representation of the instrumental equivalents of these functions.[26] The theory of the animal-machine called for systematic revision, and the mechanistic physiology of the Cartesian "man" played the role of a working hypothesis which required that every form of mechanistic explanation be specified and formulated as laws. An *a posteriori* method alone made a tentative formulation of such laws possible. But the methodological autonomy of physiology as a science is recent and comes much later than the medical empiricism developed by Thomas Sydenham in a Baconian context.

The doctrine of experience in Sydenham, as I have pointed out in *L'Empirisme de Locke,* is an historical method of classifying phenomena.[27] According to Sydenham, the knowledge of morbid entities is linked to the observation of sensible qualities which characterize their symptoms. This is the point of departure and the criterion of every attempt to classify natural realities. The method itself is based on a conception of nature which, in its essential aspects, is close to the Hippocratic tradition. Natural phenomena manifest a teleological organisation, linked in the case of pathological symptoms to conservation of health. In other words, the natural order is the principle of rationality that accounts for the development of the phenomena. The position which Sydenham takes about hypotheses relates to his scepticism about natural philosophy since *a priori* causal explanations yield only chimerical conclusions. Our understanding has a grip on reality only to the extent that it is limited to the observation of sensible phenomena and their immediate correlation in experience. Sydenham maintains that it is impossible to know the real essences of diseases. Human understanding is limited to the experimental collection of specific symptoms which reveal nothing more than the nominal essence of the diseases to be explained. Sydenham examined very carefully visible structures and changes of sensible qualities which constitute pathological symptoms, but he was indifferent to hidden organic structures. He did not believe that detailed anatomical examination and microscopic observation of elements could reveal the laws of organic functions. Research into organic forms beneath sensible surfaces and chemical explanations attempts to violate the impenetrable order of nature. They discredit the true method which is that of nature itself, which leads us by the hand in the knowledge of observable symptoms and of modes of treatment which are suggested by observation. The new disciples of Hippocrates practice a medicine which brings together an empiricism of pure observation, a belief in natural finality (that is, in divine providence which regulates the order of phenomena) and a systematic attachment to pathology and observable therapy. From this follows a denial of the possibility of reconstructing living organisms as special kinds of machines.

Physiology becomes a science in the eighteenth century with Haller, Spallanzani, Blumenbach and Bichat. This is not a direct consequence of the iatromechanism of the Malpighi kind. For the mechanistic models to which Malpighi had recourse were not sufficient to explain vital functions. Moreover, his conception of the natural order allowed unrestricted appeals to analogical explanations and even to analogical anticipations. Cartesian mechanism, which showed poor results because of its deficiencies in anatomical and "micrological" observation, was more aware of the limitations of the mechanical analogy. Sydenham's scepticism about the theory of *anatome subtilis* created a need for a new methodology which the physiologists of the eighteenth century developed. What was lacking in the seventeenth century was a conception of the organism which was biologically and experimentally useful.

Galileo's New Science of Motion

STILLMAN DRAKE

Galileo asserted that in the last two days of his *Discorsi* of 1638 he was presenting a very new science of a very old subject, motion. I see much justice in that claim. It may seem eccentric today to support Galileo's own appraisal of his work on motion and to see him as having introduced a new method, though that was the opinion of pioneer historians such as Montucla, Andres, and Whewell. Even as late as 1909, despite the doubts already raised by Raffaello Caverni, Angelo Valdarnini was still able to write:

> *Nessuno vorrà certo negare o contendere a Galileo il merito insigne d'aver usato il vero metodo sperimentale comprensivo nello studio dell'universo, e di aver esposto qua e là nelle sue Opere scientifiche le norme fondamentali del metodo sperimentale. Ma non scrisse il Galileo un apposito trattato sul Metodo Scientifico.*[1]

Thus it is only during my own lifetime that the opinion of historians has been totally reversed, following the researches of Pierre Duhem into medieval manuscripts and the studies of Alexandre Koyré which presented Galileo's work as chiefly the culmination of mathematical reasoning in a Platonic tradition, in which experiments were of no importance, and were more frequently imagined than performed. Yet often the seemingly best established conclusions have had to be modified in the light of new evidence, and the historian must be wary of taking superficial resemblances for essential continuities in the progress of ideas. Balance requires a sort of Devil's Advocate not only for Galileo, but for the great pioneers of our discipline.

It was through an excessive attention to printed books alone, and neglect of manuscripts, that pioneer historians were misled into believing that medieval physics had produced nothing of lasting value. Similarly, historians of the present era have been misled into deprecating the novelty of Galileo's science of motion through excessive attention to his final published book, and neglect of his manuscript notes. Unlike his great contemporary, Johann Kepler, Galileo did not publish an account of the bypaths and blind alleys into which he wandered before he arrived at his correct results. Instead, he published in deductive order an array of theorems and problems, stemming from a single definition and a single postulate concerning the motion of free fall. We are thus led along step by step, and we may come to feel that we are following in Galileo's own tracks, as if he had reached his conclusions in the same order as that in which he presented them to his readers many years later. And though we know how improbable it is that anyone ever worked so logically and so unfalteringly in practice, this impression has in fact led several modern historians into a certain trap.

Thus it has come about that Galileo's work on naturally accelerated motion is often treated as if he had started on it from the definition given in the *Discorsi,* and the first theorem which follows directly from it. That same theorem is then treated as indistinguishable from the Merton Rule, though no use was made of any mean-speed concept in Galileo's statement or proof; and, taking this theorem to be a legacy of the Middle Ages, some make the history of the physics of free fall appear as logical and as unfaltering as was Galileo's own final presentation of his science of motion. This simplistic view is, however, even farther from the truth than an older view, now rejected by every historian, that Galileo's science of motion began from some elementary experiments with falling bodies. Yet it has appealed so strongly to historians with a philosophical bias that some have gone so far as to question whether Galileo ever made any experiments, at least in the modern scientific sense of the word.

The simplistic historical position will not hold up, because however Galileo started on his studies of motion, he certainly did not begin with the correct definition of uniform acceleration. This is evident from his celebrated letter to Paolo Sarpi late in 1604; for no matter how we interpret the principle adopted in that letter, it was certainly not a mere legacy of the Middle Ages. Whether we take it to be a blunder on Galileo's part, or a sophisticated semantic manoeuvre to simplify the proof of a result already arrived at in another way, we cannot say that Galileo *began* his work on free fall with the correct medieval definition of uniformly difform motion or with any form of mean-speed analysis.

In the search for Galileo's procedures in reaching his new science of motion, then, we cannot rely to any large degree on the methodology he chose in presenting that science to the public. No more can we discover how Isaac Newton reached the propositions in his *Principia* from a study of the geometric proofs he offered in their support. Nor is there any real puzzle about Galileo's methodology in the *Discorsi;* its pattern is the classic procedure of Euclid and Archimedes, though supplemented by scholia and interspersed with conversational discussions of the meaning and applicability of the demonstrated conclusions. The real puzzle over Galileo's method concerns the logic of discovery, not that of proof. In the discovery of his 38 propositions on free fall and descent along inclined planes, a role was played by mathematics, another role was played by experiment, and still another was played by chance, or luck. Errors were made and corrected; wrong leads were followed and abandoned. To reconstruct the historical process we must turn to other documents—to treatises Galileo withheld from publication, to his letters, and above all to his notes and working papers. From these we may discover Galileo's methodology in the science of motion; or rather, we can reconstruct it, for it was doubtless largely unconscious on his part in the early stages, and unconscious methodology is perhaps a contradiction in terms.

The task of reconstruction is difficult, but not impossible. It is difficult because Galileo's surviving notes are undated, and were intended only for his own eyes. But it is not impossible, because the notes of any one man must have a chronological sequence, and those of a man like Galileo must also have had a psychologically plausible, if not a logically necessary, concatenation. The problem, then, is to find an order for these seemingly chaotic notes that is internally plausible, and is also consistent with everything else we know about Galileo and his work.

Fortunately, some benchmarks are provided by dated documents or by others that can be placed in time with reasonable accuracy. One of these, of course, consists of the lectures *De motu* that certainly belong to the Pisan period, and probably to the year 1590. Another is Galileo's syllabus on mechanics, probably brought to its final form around 1600. Certain letters, notably in 1602, 1604, 1609, and 1611, also give us definitive evidence about Galileo's state of knowledge regarding particular problems of motion in those years.

Characteristic forms of handwriting at various times also provide clues, as do the watermarks in the paper used for the notes. Clues from watermarks are particularly valuable because the papers Galileo used at Pisa, at Padua, and at Florence came from different makers; and in a few cases, the same paper was used for dated letters and for notes on motion. Copies made by pupil-assistants at Florence of theorems proved earlier at Padua establish still further details of order.

Employing these kinds of evidence, I have been able to reconstruct a pattern of Galileo's progress in the study of motion which, while always subject to correction and refinement, seems to me likely now to hold up well in its main outlines. From this I shall select representative procedures followed by Galileo in reaching his principal results.

Galileo's demonstration of equilibrium conditions on inclined planes in *De motu,* by reduction to the principle of the lever, led him to a proof that a body should be moved on the horizontal plane by a force smaller than any previously assigned force, and he noted that such motions should be called "neutral" rather than natural or forced. In attempting to deal with *motion* on inclined planes, however, he assumed that speeds should be proportional to effective weights, on the Aristotelian basis that weight is the cause of motion. From this it followed that the speeds of a given body along two planes of equal height should be inversely proportional to the lengths of the planes. Such ratios, Galileo noted, were not borne out by experiment, which was perhaps the reason that his treatise *De motu* was ultimately withheld from publication.[2]

In 1602, Galileo wrote to Guidobaldo del Monte in answer to the latter's objection to his statement that the times of fall along chords to the bottom of a vertical circle were equal.[3] He declared that he had a proof of this proposition, and also of another, that the time of fall was longer along a chord than along its two conjugate chords, but said that he had been unable to prove the isochronism of descents along circular arcs, which was his objective. Since Galileo did not yet have the law of free fall in 1602, it is reasonable to ask the nature of his proofs concerning chords. These appear not to survive, but in fact one can rigorously prove the equality of times along chords to the bottom of a vertical circle from the earlier, incorrect, assumption that speeds along different planes of the same height are inverse to the lengths of the planes.[4] Since only overall speeds are concerned, acceleration need not be considered. The shorter time along conjugate chords follows also from this proof. These two theorems are easily confirmed by experiment, by placing boards of different length in a hoop, and noting whether balls released simultaneously along them reach the bottom together. Quite possibly, Galileo discovered this interesting theorem in that way, and then found a proof from his old false assumption.

In the same letter, Galileo described the use of long pendulums in the study of descent along arcs of vertical circles. Such observations probably made him question his idea in *De motu* that acceleration is only of brief duration, at the very beginning of motion. At any rate, by mid–1604 at the latest, he had found a law relating ratios of speeds in descent to the distances from rest, finding that the acquired speeds when plotted against distances fell along a parabola. On the discovery document, *f. 152r* of volume 72, among the manuscripts preserved at Florence, Galileo started with the assumption that as the distances increase as the natural numbers, overall speeds grow as the triangular numbers, a rule suggested by medieval sources which I will discuss later. But Galileo found that for continuous growth, this led to contradictory speed ratios, and he accordingly amended it by introducing a mean proportional of distances from rest, and by using this in the ratios of time.[5] Thus the times-squared law emerged from an arbitrary mathematical device intended to reconcile conflicting ratios of speeds.

Galileo's next move, naturally enough, was to test whether the new rule held good also for inclined planes, and this again was a purely mathematical exercise. The sheet on which it was carried out, *f. 189r,* implies all but one of the rules he would need for the comparison of motion along inclined planes in general, from rest or after a given fall, though this was not immediately apparent to him.[6] On this same sheet there is a sketch of the parabola mentioned in the last line of the previous discovery, and there is also another parabola converted into a triangle by extending the abscissae. Again an arbitrary mathematical device—this time, that of squaring—enabled Galileo to simplify his

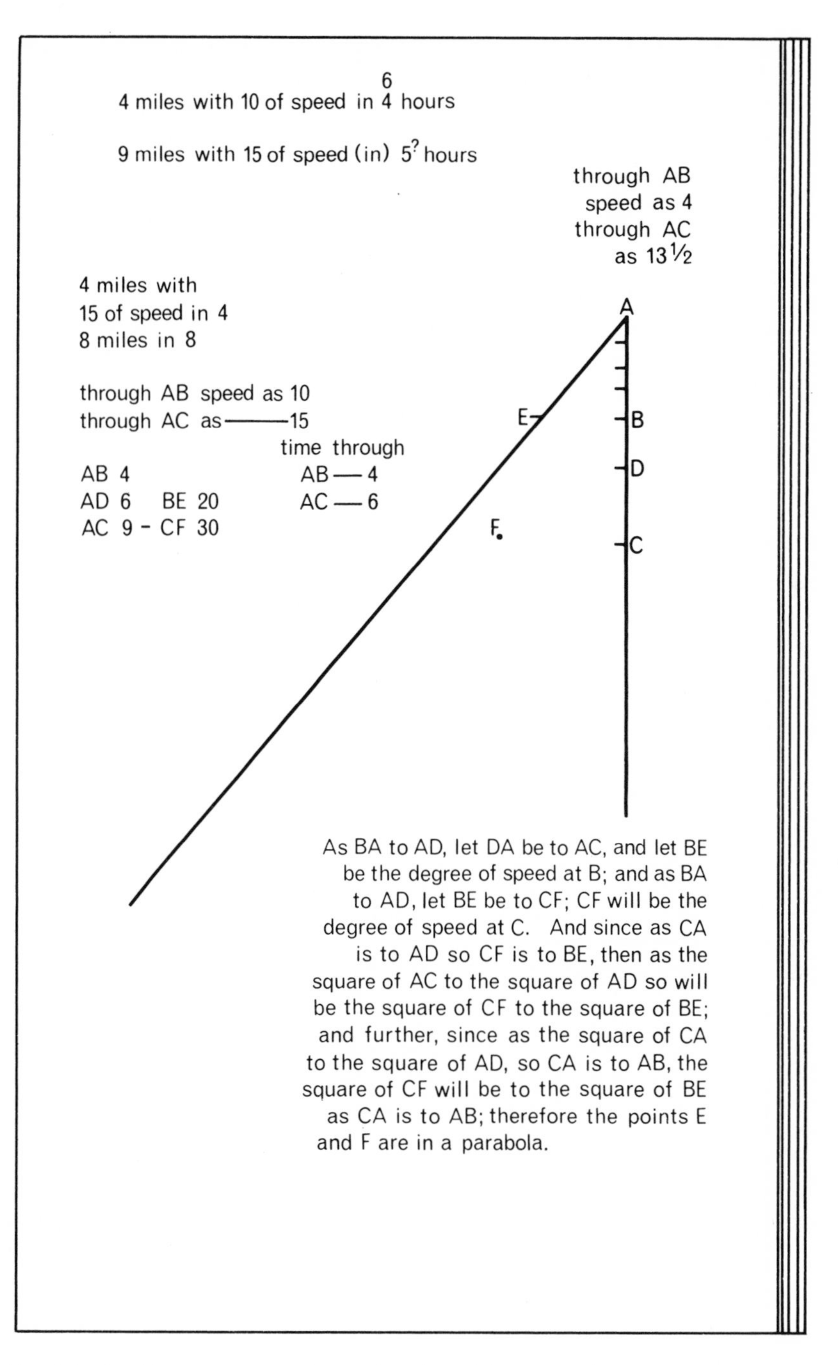

6
4 miles with 10 of speed in 4 hours
9 miles with 15 of speed (in) 5? hours
through AB
speed as 4
through AC
as 13½
4 miles with
15 of speed in 4
8 miles in 8
through AB speed as 10
through AC as——15
time through
AB 4
AD 6 BE 20
AC 9 - CF 30
AB — 4
AC — 6
A
E
B
D
F
C
As BA to AD, let DA be to AC, and let BE be the degree of speed at B; and as BA to AD, let BE be to CF; CF will be the degree of speed at C. And since as CA is to AD so CF is to BE, then as the square of AC to the square of AD so will be the square of CF to the square of BE; and further, since as the square of CA to the square of AD, so CA is to AB, the square of CF will be to the square of BE as CA is to AB; therefore the points E and F are in a parabola.

Partial English transcription of folio 152*r*, Vol. 72, *MSS Galileiani, Biblioteca Nazionale di Firenze,* showing Galileo's steps in the discovery of the law of free fall, 1603–1604.

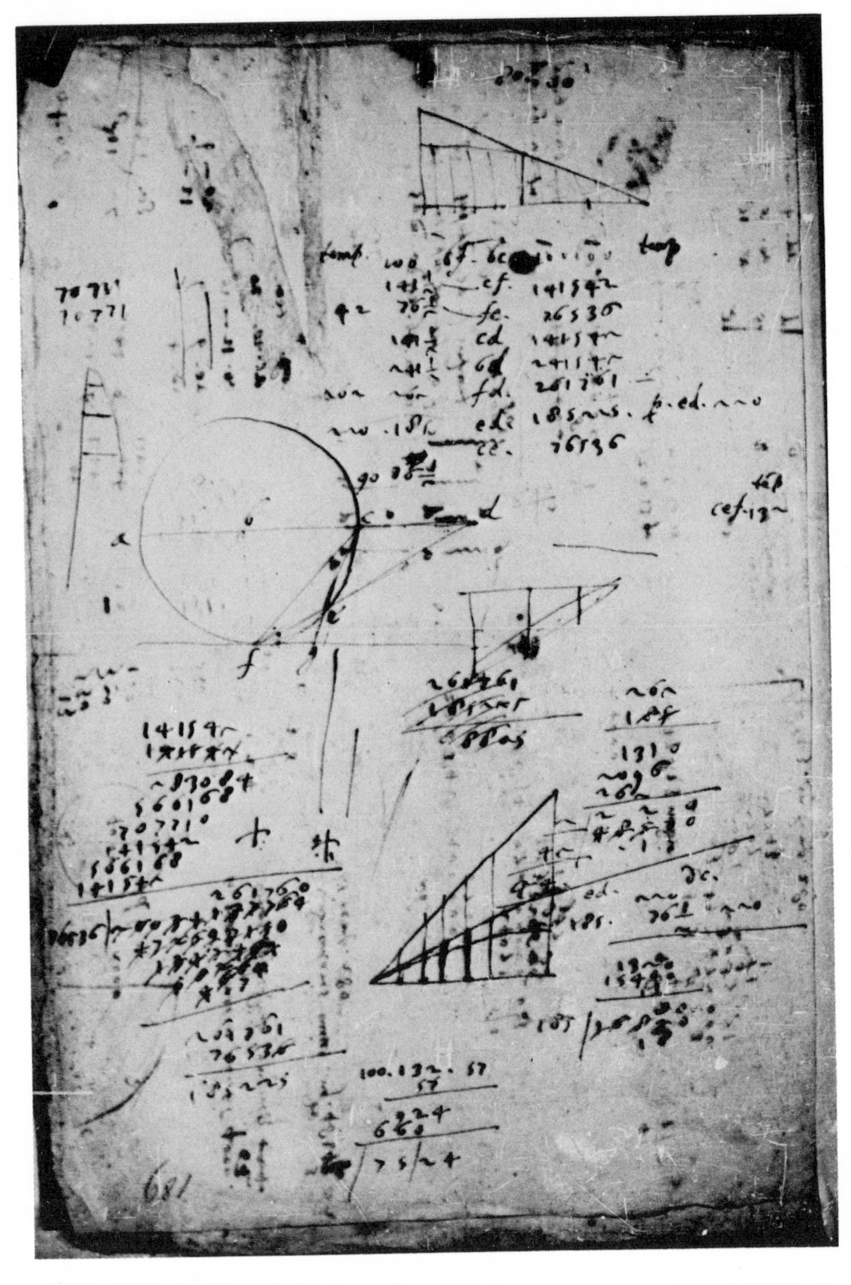

Folio 189*r* (from Vol. 72, *MSS Galileiani; courtesy of the Biblioteca Nazionale di Firenze*).

Folio 175 (from Vol. 72, *MSS Galileiani; courtesy of the Biblioteca Nazionale di Firenze*).

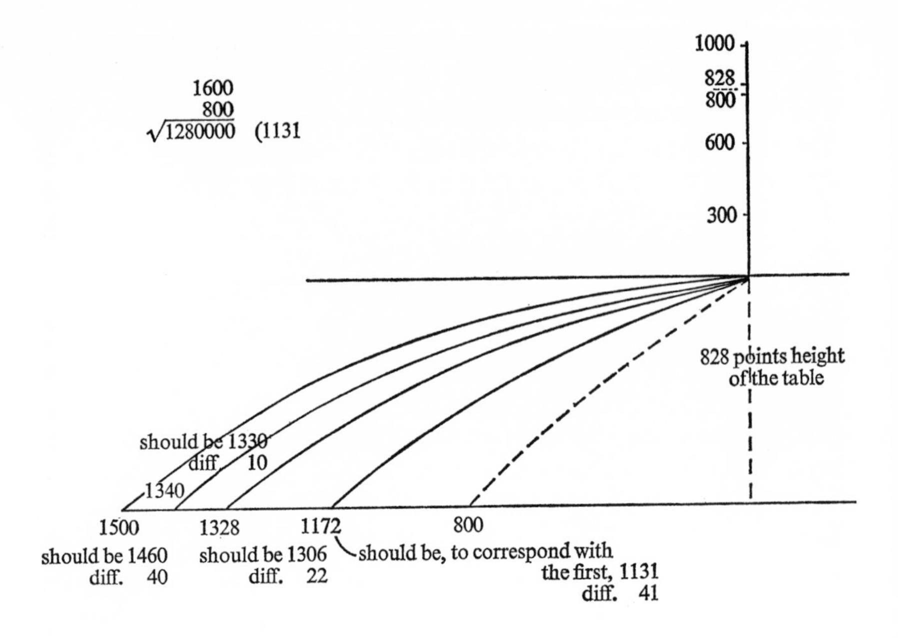

[*partial calculations omitted as not used*]

300. 828.
800
300 /662400 (2208

300. 800
800
300 /640000 (2133

2208
800
√1766400 (1329

2133
800
√1706400 (1306

300. 1000
800
300 /800000 (2666

2667
800
√2133600 (1460

[*partial calculations omitted as not used*]

English transcription of folio 116^{v}, showing Galileo's calculations of horizontal distances expected under his mean-proportional rule, using shortest drop as basis. Unused partial calculations are omitted as are trial divisors in root extraction and related remainders. (*Courtesy of Isis*) procedure by adopting a new definition. This particular conceptual change, however, was destined to produce a semantic obstacle for historians, who had only the demonstration that resulted and not this unpublished preliminary step. When Galileo wrote to Paolo Sarpi in October, 1604, and drew up a proof for him, he did not consider the speeds acquired as squared, but simply adopted the squaring as part of the meaning of the word *velocità,* which had not previously been assigned any physical definition. Thus, he offered the rule that *velocità* were proportional to distances from rest, adducing experience of machines that act by striking as his example. His wording in 1604 thus made it seem that Galileo literally asserted a rule of proportionality of speeds acquired to distances fallen, in the sense of those words today, though there is no instance in any of his notes of application of that rule, even in the years 1604–09. Early in 1609 he adopted our ordinary meaning of "velocity"; meanwhile he had worked out most of the essential theorems found in the Third Day of the *Discorsi.*

With the law of free fall in hand and mathematically extended to descent along inclined planes, where it could be verified experimentally, Galileo returned to his earlier project of seeking a rule for times along arcs of a vertical circle. This time his method was one of detailed numerical calculation. Starting with the chord of a quadrant, on f. 166*r,* Galileo drew the two equal conjugate chords, and then the four, and the eight, obtained by successive bisections.[7] He proceeded to compute the times in straight and broken fall along these, in various combinations. In so doing, he arrived at the calculational technique that was needed for the comparison of times along any set of slopes, beginning at any height. But after writing some vain conjectures about circular fall across his elaborate diagram, Galileo seems to have lost interest in this quest. Instead, he decided to write out and prove theorems about motion in the vertical, on inclined planes, and along broken lines. These notes, made at Padua in 1607–08, have been published by Favaro.[8] The work at this time was entirely geometrical, with relatively few numerical examples and none that suggest experimental data.

Toward the end of 1608, however, there is clear evidence of Galileo's use of precise experiment in the modern scientific sense. In the autumn of that year, it occurred to him that his mathematics of free fall had put into his hands a means of testing his longstanding belief that a body in horizontal motion would continue uniformly if unimpeded by external resistances. This belief had originated with his proof in 1590 that a body on a horizontal plane should be capable of being set in motion by a vanishingly small force. In his *Mechanics,* about 1600, the same idea had been further developed, and by 1607 he was teaching that though force was necessary to commence motion, absence of resistance sufficed to conserve it.[9] Contrary to Peripatetic tradition, but in accordance with ideas in the ancient pseudo-Aristotelian *Questions of Mechanics,* Galileo held that different tendencies to motion could exist in the same body, and that these would give rise to a single resultant motion.[10]

Given all this, Galileo had needed only some way of knowing the ratios of speeds of a body leaving a horizontal surface and falling freely through a fixed distance. Those ratios could now be known by dropping a body from rest through various heights and then deflecting it horizontally. The apparatus is drawn on f. 175*v,* and the experiment is recorded on f. 116*v.* A table 828 *punti,* or about 80 cm. high, was used, and horizontal travel during this drop was recorded for initial drops to the table of 300, 600, 800, and 1000 *punti.* Using the initial drop and its related horizontal advance as a standard, Galileo computed the others, which were in very good agreement with the actual measured distances of advance during fall.[11]

This part of the experiment alone marks a methodological epoch in the study of motion. All known earlier studies of motion (including Galileo's) appear to have been based either on more or less casual, even if careful, observations, or else on pure reasoning, logical or mathematical, from explicit or implicit definitions. This appears to have been true of ancient studies, including those of Aristotle and of Archimedes, as well as of medieval studies by Bradwardine, the "calculators", and Buridan. Galileo's own *De motu* indicates that he had tested some theoretical ratios of speed, but that they had not been vindicated by experiment. The present case is rather different. Ratios of speeds that had alreay been tested and found to conform to experience were now applied to a physical theory of compound motion in order to test a physical theory about one of its component motions. The test was successful, meaning that more could safely be built on these theories.

But this was not all. Galileo next added a drop of 828 *punti* to the table, equal to the drop after horizontal deflection. The purpose was evidently to confirm his double-distance rule, derived in 1607 by reasoning from one-to-one correspondence of uniform and uniformly accelerated speeds.[12] But in this test, the horizontal travel fell far short of what Galileo expected. The reason was that he did not know of the factor of 5/7 that applies to the acceleration of a rolling body as against one falling freely. Galileo had no way of knowing this, so he attributed the loss of motion to the impact at the moment of deflection. Accordingly, he devised a new experiment, shown on f. 114*v*, in which the ball simply rolled off the end of an inclined plane. His recorded data permit reconstruction of the entire experiment, and though he was unable at the time to compute the

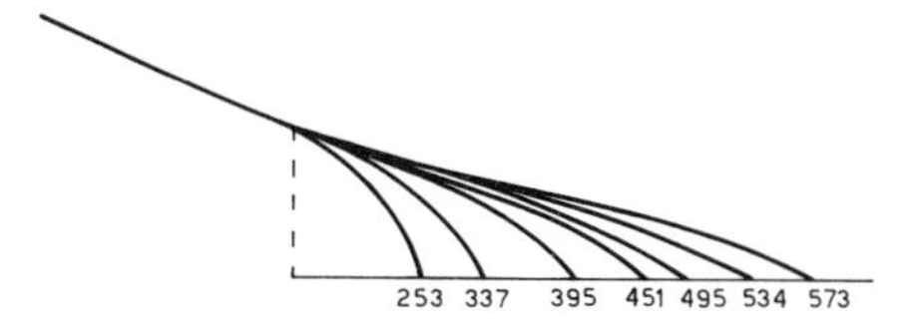

Diagram on folio 114 (from Vol. 72, *MSS Galileiani, Biblioteca Nazionale di Firenze*).

horizontal progress for an oblique impetus, we can do so. The result shows him to have been remarkably accurate in recording his results. Using 980 cm/sec^2 for the value of acceleration in free fall, and 5/7 of this for rolling, the greatest discrepancy from theoretical values is about 3%. Descents along the slope were 200, 400, 600, 900, 1200, 1600, and 2000 *punti,* the final drop for a 30° plane being 500 *punti.* The devising of such an experiment is itself methodologically significant.

The accuracy of these recorded data suggest that Galileo had a good deal of practice as an experimentalist before 1608. This contradicts the prevailing opinion, drawn only from Galileo's printed books and our knowledge of the customary practices of physicists up to his time. Even more interesting is the care that Galileo took to assure himself of the validity of his basic assumptions, in view of the fact that he did not later describe these experiments for his readers.

There is still another aspect of this work that is of methodological importance, and that is that Galileo's experiments in 1608 led on immediately to a new discovery. The height of the table was such as to permit him to stoop after releasing the ball, and observe its path from table to floor. These paths he drew in on f. 116, with fair accuracy; and he noticed them to be approximately parabolic. On f. 117*r,* which bears the same watermark and which contains notes of a preliminary observation that led to the experiment, we find drawings of parabolas and of a parabolic trajectory with its tangent. Galileo turned at once to the demonstration of the parabolic trajectory, on various sheets that can be dated early in 1609 and which are associated with a letter written early in February of that year. It was in relation to this analysis that Galileo finally adopted the physical definition of "velocity" that we still use. His reasoning is embodied in the third proposition of the Fourth Day of the *Discorsi,* which is virtually unchanged from the form in which it was written early in 1609.[13]

I have given but a sketchy account of the chronology and nature of Galileo's early work on motion, up to the point at which the main body of propositions had been reached that were later to appear in the *Discorsi.* However elementary the content, the pattern is that of modern scientific investigation. First, a causal hypothesis was adopted that turned out to be wrong; namely, that speed of fall depends on effective weight. The incorrect ratios derived for speeds on inclined planes were tested, found defective, and so reported. *De motu* was withheld from publication, but the problem was not forgotten. In time, two theorems were found that agreed with experience and that could be proved from the false assumption. This time the circumstances of testing drew attention to the role of acceleration, previously neglected. Search for a rule of acceleration revealed contradictions in the traditional law. A mathematical device, introduced to reconcile ratios of speed in continuous acceleration, led to the law of free fall. This law was found consistent also with descent on inclined planes, but in a form inconvenient for relating speeds to simple distances from rest. Accordingly, speed was redefined in a manner consistent with observations of a different kind. There followed a long train of mathematical theorems, in which the concept of speed seldom appeared, being replaced by relations of distances and times. The new law permitted precise testing of an earlier physical hypothesis related to inertial motion. This test succeeded, and led incidentally to a new discovery. Mathematical development of the new phenomenon required a re-definition of speed. At this point, Galileo's new science of motion was completed in all essential respects except the decision as to the best postulational basis for its formal deductive presentation. If none of the elements in this pattern are new—causal hypothesis, mathematical expression, experimental test, new discovery, reconciliation of concepts by new definition, and so on—these elements were nevertheless combined in a manner for which it is hard to find earlier examples in the study of motion, and perhaps in any area of physics.

Nevertheless, Galileo did not disdain to borrow from earlier sources anything that seemed to him to fit in with his conception of exact science. Most notable is his adoption, as his first theorem on uniform motion, of the first proposition of Archimedes in his treatise *On spiral lines.* This theorem is of capital importance with regard to the principal respect in which Galileo's science of motion departed mathematically from medieval physics. Its proof by Archimedes depended solely on the Eudoxian theory of proportion, which was lost in the standard Latin text of Euclid during the Middle Ages and was restored only in the sixteenth century. It was that theory of proportion alone that permitted Galileo to deal with acceleration as continuous in the modern mathematical sense, the importance of which fact I shall presently stress. Borrowing also Aristotle's unchallenged definitions of "equal speed" and "greater speed", Galileo applied the same Eudoxian definition to the proof of his second theorem on uniform motion. This appears to be novel, as do his other four theorems which ingeniously used the notion of compound ratio also used by Archimedes in lieu of the later concept of "function".

From medieval tradition, Galileo at first adopted a rule of free fall related to impetus theory, though he found this inconsistent with continuous acceleration and abandoned it, as I have already indicated. It is also possible that he found in Swineshead's "Calculations" the first clue to his own fruitful use of one-to-one correspondence betwen infinite aggregates, in its applications to proportions, though this is dubious on other grounds. It is highly improbable that either Galileo's method or his results in the science of motion are *founded* on medieval mathematical physics, for reasons I shall now set forth.

Antiquity offers no mathematical treatment of free fall or of uniform acceleration, but writers of the fourteenth century afford both. The odd thing, in the light of the high quality of both the mathematics and the physics of that century, is that free fall and uniform acceleration were then dealt with quite separately, and not in conjunction. I shall try to show that our surprise at this has arisen from a basic oversight on our part, and from a misinterpretation of medieval physics as a whole that has arisen in the process of singling out certain parts of it precisely because, in the light of later knowledge, they seemed to be harbingers of Galileo's science of motion. But in fact those parts not only belonged to a different physics, founded on the search for causes, but also depended on a different mathematics, essentially arithmetical and discrete rather than geometrical and continuous.

First, let us consider the medieval treatment of uniform acceleration. This arose in the classification of every possible type of motion, as Father William Wallace has shown, rather than in the quest for a mathematics of free fall.[14] It is no accident that Galileo's treatise in the *Discorsi* is entitled "On naturally accelerated motion" and not, in the medieval pattern "On uniform acceleration". Galileo's purpose here, as elsewhere in science, was to limit the scope of his inquiries to separate and well-defined areas, and not to seek a general theory of the universe. This is an extremely important part of his scientific methodology, expressed in his *Dialogue* when he wrote that there is no phenomenon in nature, not even the least that exists, which is such that even the most profound theorist will ever attain a complete understanding of it. Earlier, in the *Assayer,* he had remarked that the best philosophy will offer the fewest promises and will teach very little as certain.

The concept of a mean speed as a single speed chosen from an infinite aggregate to represent all speeds, so important to medieval analysts of acceleration, is nowhere to be found in Galileo's writings, or even in his private notes. Nor was the mean proportional, so essential to Galileo's treatment, given any particular meaning in the Middle Ages. Medieval physics dealt always with bounded motions, which implied a mean speed. Galileo considered unbounded motions, within which ratios may be found between speeds, times, or distances at any two points, but for which the concept of a mean speed is superfluous. It is noteworthy that even Galileo's earliest triangles for accelerated motion were open at the bottom, as was that used in the proof of the times-squared law in the *Discorsi* near the end of his life. It is likewise noteworthy that Oresme, who first used the triangle for acceleration, implied that the area represented total motion, or distance traversed, whereas Galileo never implied any meaning for such an area other than that of a kind of overall speed. These conceptual differences are related to the medieval search for causes as against the quest for laws, and to the assumption of discreteness as against that of continuity in physical phenomena.

Next, as to the medieval treatment of free fall as such, which was a part of Buridan's impetus theory, I was formerly inclined to agree with Alexandre Koyré that Galileo, in his *De motu,* simply adopted bodily the impetus theory of the fourteenth century, and that he subsequently abandoned it only because it did not lend itself to precise mathematical formulation. This, however, is not so. The very ingenious explanation of acceleration in free fall offered by Jean Buridan was not even mentioned in *De motu,* though it had recently been expounded again by Benedetti. In its place, Galileo offered an inferior theory, so it cannot be maintained that he adopted impetus theory bodily. But more than that, it cannot be maintained that Buridan's impetus theory of free fall resisted precise mathematical formulation. Such a formulation had in fact been given to it by Albert of Saxony, and this was accepted by Oresme, by Leonardo da Vinci, and probably by all physicists up to the second half of the sixteenth century. Galileo himself attempted to use it when he first seriously sought a rule of acceleration in free fall, probably early in 1604. Albert's formula, I believe, has simply been misunderstood by modern historians, through neglect of medieval physics as a whole. To understand it, let us review the physics of impetus.

Buridan's explanation of acceleration in free fall was the first fully rational account ever to be given, and the best to appear up to the time of Newton. It is perhaps still the best explanation for those who insist on causes, and are not content with laws. Buridan regarded impetus as a kind of force impressed in a body and remaining there except as reduced either by external resistances or by an internal tendency on the part of the body to some contrary motion. In the case of the heavenly spheres, neither external resistence nor tendency to any motion except rotation was present, whence an initial impetus given to them would be conserved forever. In terrestrial heavy bodies, however, there was always a natural tendency downward, which tendency would normally conflict with and reduce the impressed impetus, and ultimately would bring the body to earth.

Buridan, however, was an excellent physicist, and he did not fail to see that there was one unique case in which impetus would be conserved undiminished in a heavy body, in the absence of any external resistance. This was the case of a heavy body hurled straight down, the natural tendency being also in that direction. For the explanation of acceleration in free fall, only a single additional postulate was required; namely, that impetus could be imparted not only by violent projection, but also by natural motion itself. And thus Buridan wrote that one must imagine that in the first movement from rest, a body is moved only by its heaviness; but along with that motion, it gains an impetus, so that in the next movement it is moved by that impetus together with its heaviness, and hence moves faster. With the faster motion it receives still stronger impetus, and so on, the speed increasing to the end.[15]

Under such a theory it is evident that during the first movement from rest, when only a single cause—the heaviness—is acting, the speed is uniform; and in the second movement it is likewise uniform, though faster, since only the first impetus is acting along with the heaviness, and so on. Buridan's explanation in fact implies a sort of quantum theory of speeds, increasing discretely though very rapidly. This accords well with Aristotelian causal principles, as well as with Aristotle's own remark that though the motion of projectiles appears continuous, it is not really so, only that motion which is imparted by an unmoved mover being truly continuous. Gassendi was later to support a theory of successive discrete impulses in his book entitled *On the Motion Impressed by a Moved Mover,* and I shall presently consider another post-Galilean impetus theory. In the Middle Ages, a quantum physics of acceleration was also mathematically appropriate because (as previously mentioned) the theory of proportion was arithmetical and discrete, in the absence of one Eudoxian definition in the medieval Euclid.

In strict Aristotelian physics, there could be no first *mathematical* instant of motion from rest, for at such an instant a body would be both in motion and at rest, in violation of the law of contradiction. But there could be a first *physical* instant of motion, meaning by this a duration shorter than any previously assigned time. In short, the successive "degrees" of speed, which were always represented by integers, were contiguous rather than continuous.

With this in mind, let us now consider the mathematization of Buridan's theory of free fall by Albert of Saxony. Albert's rule stated that when a body has fallen a certain distance, it has a certain speed; and when it has moved twice that distance, it has twice that speed, and when it has gone triple the first distance, it has triple the first speed, and so on. Historians of our time, thinking in post-Galilean and algebraic terms, have naturally supposed this to mean that the distances of which Albert spoke were to be taken *from rest,* in which case the rule would amount to putting speeds acquired proportional to distances from rest. Such a rule was indeed put forth in 1584 by Michel Varro, and it was shown by Galileo to be untenable and to imply an instantaneous motion.[16] Indeed, this rule contradicts the concept of rest itself. But Albert of Saxony would have been perfectly capable of perceiving that contradiction, and his rule was in my opinion quite different. The distances of which he spoke were meant to be taken successively, so that in the second distance the body went twice as far and twice as fast as in the first, implying an equal time, and so on. Thus in Albert's rule, the successive distances, speeds, and times, all went up as the natural numbers, (and any cumulative values progressed as the triangular numbers). It was this rule that was accepted by Oresme, Leonardo, and others down to the time of Galileo, who indeed tried to apply it in the opening lines of f. 152*r*, but mistakenly. He found that it could not apply to continuous acceleration, and moved on to the times-squared law. But there was nothing internally contradictory in Albert's rule; in an Aristotelian world in which bodies fell with different speeds, proportional to their weights, a rule of this kind might well apply.

This account of Albert's mathematical rule for fall clears up an otherwise perplexing problem of historians; that is, why medieval writers did not associate free fall with uniformly difform motion. In the latter, it was known that the distance traversed in the second of two equal times from rest must be three times that covered in the first. But Albert's rule for free fall would make the distance in the second time only double that of the first; hence free fall could not be a case of uniform acceleration. This account also clears up a second puzzle; namely, why in all the commentaries on physics after Albert, no one is known to have raised the question whether, in free fall, the speeds are proportional to the distances fallen or to the times elapsed. In the usual modern interpretation of Albert's rule, such a question would have been bound to arise long before Galileo.

Any lingering doubts concerning this interpretation of impetus mathematics are dispelled if we consider a suggestion of G. B. Baliani in 1646 that behind Galileo's odd-number rule for distances in equal times, as observed by experiment, lay the natural-number rule for vanishingly small times. Baliani noted that if we take three observable equal times, and the distances traversed, and if we divide each of these distances into ten parts in which the spaces increase as the integers, then 55 such tiny units of space are covered in the first large equal time, 155 in the second, and 255 in the third; but ignoring the last digit, these distance-quanta are 5:15:25 or as 1:3:5, very nearly. Now, if we had made the division by one hundred instead of ten, said Baliani, the figures would have been 5050, 15050, 25050, very close to 1:3:5; and since physical instants are millions of times smaller than observable times, Galileo's odd-number rule for large distances is implied by a natural-number law for infinitesimal distances.[17]

In the same year Honoré Fabri published at Lyons a very detailed treatise on the impetus theory of free fall, in which similar ideas were developed at length. Among other things, Fabri argued against Galileo's position that in order to reach any speed from rest, a body must have passed through every possible smaller speed, a position also rejected by Descartes.[18] Fabri noted that the same body, falling freely, and along an inclined plane, goes faster vertically. Hence, he said, it must start faster; and therefore there must exist some speeds along the incline that are not to be found in vertical fall. Such arguments help us to understand medieval impetus theory, with its implicit assumption of infinitesimal successive motions, as against Galileo's continuity theory of motion, made possible by the restoration of Euclid, Book V. Not only medieval physical thought, but the elaborate medieval arithmetical theory of proportion continued to be taught well into the seventeenth century, and conservative scholars were not swept away by Galileo's new science of motion.

Even more fundamental was Fabri's contention that his physics was better than Galileo's because it could provide a cause for acceleration in fall, while Galileo's could not. Here we have a truly methodological debate, though Galileo was not alive to reply. His view was that whatever laws were confirmed by precise experiment should be assumed to hold beyond the boundaries of observable phenomena, much as Newton was to give it as a rule of science that properties found in all bodies accessible to experience were to be attributed to all bodies whatever. Ultimately, this procedure would eliminate causes in favor of laws, and we all know what Galileo said about inquiries into the cause of acceleration in fall. Whether or not we happen to agree with the substitution of laws for causes as a program of physics, it will be admitted that such a program had not been suggested before Galileo, and that it has received a good deal of support since his time. Fabri argued, on the other hand, that his physics and Galileo's were both consistent with the observable phenomena, while only his could discover something ultimate behind those phenomena, and different in form from their laws.[19]

In this sense there is perhaps something methodologically new even in Galileo's ultimate presentation of his science of motion to his readers. Previously, I outlined the novelties in Galileo's procedures in reaching his conclusions, but said that his presentation of them was simply that of Archimedes, interspersed with scholia and interpretative conversations. I am not unwilling to stand by that, and it was certainly Galileo's own view, expressed in a letter which

I shall cite presently. But it is worth mentioning that if that very form of presentation was not an open and explicit attack on the Aristotelian conception of physics as an understanding of nature in terms of causes, it at least implied that everything of lasting value in physics could be presented in the form of precise laws, experimentally confirmed, and their deductive consequences. It is my view that Galileo's reason for presenting part of his last work in Latin, within an Italian dialogue, was to mark out those parts which he himself considered to be established irrefutably, and to which he wished his own name, rather than that of any interlocutor, to be forever associated. And it was precisely his new science of motion, consisting of definitions, postulates, and theorems, that was published in Latin, for scholars everywhere. Not even the theorems in his new science of strength of materials, which remained to some degree tentative, were so distinguished from the discursive and speculative parts of his last great book.

Those who look upon Galileo's science of motion as in effect the mere addition of some new results to medieval physics, carried on logically from earlier beginnings should, I think, explain historically why the times-squared law and the odd-number rule were not deduced much earlier and applied to inclined planes and projectile motions. One common explanation is a fancied inhibition of scientific thought by the rise of humanism. Now, no intellectual movement has ever been so strong as to seduce every individual thinker for two centuries away from all other pursuits, so this is not so much an explanation as an *ad hoc* reason—what Galileo called "reaching for sky-hooks". But in fact it is so far from the truth that the very opposite is true. It was a humanist, Bartolomeo Zamberti, who, moved by his abhorrence of everything touched by the Arabs and his Renaissance love for Greek antiquity, first made accessible in print a correct Latin translation of Euclid's fifth book. But Zamberti did not explain its true significance. The historical event that made possible Galileo's new science of continuous acceleration in free fall was therefore not Zamberti's Latin Euclid of 1505, but Nicolò Tartaglia's Italian Euclid of 1543, in which the real meaning of Eudoxian proportion theory was explained, and the specific correction of the errors in the medieval commentary of Campanus was set forth. But since the universities did not use Italian texts, and especially such colloquial Italian as Tartaglia's, intended to educate the layman, it was not until the 1570's that Eudoxian proportion theory began to invade the universities. The Latin commentaries of Federico Commandino and of Christopher Clavius date from that decade; and thus Galileo, who entered the University of Pisa in 1581, was among the first generation of students who could reasonably be expected to put aside the medieval theory of proportion and to think in terms of ratios of continuous magnitudes in the modern sense. And by the time Galileo published his work, there was a new generation of mathematicians capable of understanding and advancing it, though there were also still men like Fabri and Descartes to oppose his new science of motion. I mention this because it seems to me that datable events are preferable to intellectual and social theories in the explanation of new modes of thinking in physical science.

In conclusion, here is Galileo's own account of the methodology used in his ultimate publication, written in a letter to Baliani in 1639:

> I have treated the same material [as you], but somewhat more at length and with a different attack; for I assume nothing but the definition of that motion with which I wish to deal, and whose events I wish to demonstrate, in this imitating Archimedes in his *Spiral lines,* where he, having explained what he means by motion made in a spiral—that is, that it is composed of two equable motions, one straight and the other circular—goes on immediately to demonstrate its properties. I explain that I wish to examine the properties that are found in the motion of a moveable which, leaving from a state of rest, goes moving with speed always growing in the same way;

> that is, that the acquisitions of speed grow not by jumps, but equably with the growth of time, so that the degree of speed acquired, for example, in two minutes of time, shall be double that acquired in one minute, and that acquired in three minutes, and then in four, is triple and then quadruple that which was acquired in the first minute. And premissing nothing more, I go on to the first demonstration in which I prove that the spaces passed by such a moveable are in the squared ratio of the times; and I go on to demonstrate a goodly number of other events. You touch on some of these, but I add many more—and perhaps more marvellous [ones], as you will see from my dialogue on this matter, already published two years ago at Amsterdam, though none have come to me except page by page, sent for corrections and for the making of an index of important matters . . .
>
> But getting back to my treatise on motion, I argue *ex suppositione* about motion, so that even though the consequences should not correspond to the events of the natural motion of falling heavy bodies, it would little matter to me, just as it derogates nothing from the demonstrations of Archimedes that no moveable is found in nature that moves along spiral lines. But in this I have been, as I shall say, lucky; for the motion of heavy bodies and its events correspond punctually to the events demonstrated by me from the motion I defined. . . .[20]

From this, it seems to me that Galileo's own view of his method in the science of motion was that which was expressed only much later, by Heinrich Hertz, in these words:

> We form for ourselves images or symbols of external objects, and the form which we give them is such that the necessary consequents of the images in thought are the necessary consequents in nature of the things pictured. In order that the requirement be satisfied, there must be a certain conformity between nature and thought. Experience teaches us that the requirement can be satisfied, and hence that such a conformity does exist.[21]

Sources of Galileo's Early Natural Philosophy

A. C. CROMBIE

"He exalted Plato to the skies for his truly golden eloquence, and for his method of writing and composing in dialogues; but above everyone else he praised Pythagoras for his way of philosophizing, but in genius he said that Archimedes had surpassed all, and he called him his master". The omission of Aristotle's name from this honours list by Galileo's second seventeenth century biographer, Niccolò Gherardini, is no surprise; nor is his preceding remark that, far from following current fashion in running Aristotle down, Galileo praised his marvellous writing on literature and ethics but found that "this great man's way of philosophizing did not satisfy him, and that there were in it fallacies and errors" (Galileo, *Opere,* xix, 645). Nevertheless, I shall respond to the invitation given to me to discuss briefly some "wider issues" relating to Stillman Drake's very interesting paper, by taking up just one question on which I shall argue that Aristotle had a far more profound influence on Galileo's scientific thinking than remarks such as Gherardini's might suggest.

Professor Drake makes a point of stressing Galileo's alleged decision "to limit the scope of his inquiries to separate and well-defined areas, and not to seek a general theory of the universe". He seems to refer to the range of content or subjects Galileo was prepared to consider. But going on to say that this is "an extremely important part of his scientific methodology", he cites the *Dialogo* and *Il Saggiatore* for examples of Galileo's limit being placed on the expectation of certainty rather than the range. Galileo's performance in scientific inquiry was undoubtedly guided by his policy of selecting acceptably answerable questions as much as by his criteria for acceptable answers. But whether Professor Drake means that Galileo limited the range or the certainty he expected science ultimately to achieve, I should argue that the opposite is true.

First, Galileo's very effective method of limiting problems in order to solve them was nearly always aimed in the end, whether through the science of motion and mechanics or through telescopy, precisely at establishing not only true methods of natural philosophy, but also the true general theory of nature. This was a theory comprising matter and its properties as discovered by both terrestrial and celestial inquiries, their bearing on cosmology, the relation of perceiver to perceived and of knower to known, and the bearing of it all on theology.

Secondly, throughout his scientific inquiries and debates, Galileo wrote continually of finding "true and necessary demonstrations" (*Opere,* ii, 155; v, 330) of his conclusions, and on one famous occasion, in his First Letter about the Sunspots (1612), he looked forward not un-typically to solving "the greatest and most admirable problem there is, the true constitution of the universe. For such a constitution exists, and exists in only one, true, real way, that could not possibly be otherwise" (*Opere,* v, 102). Strong words; in fact, the words of Aristotle's *Posterior Analytics* (i.2, 71b9–72a24; 6, 74b5–6; 10, 76a31–b31), well known in Galileo's day to every educated person. We have unqualified scientific knowledge of something, Aristotle had written, when "we know the cause on which the fact depends, as the cause of that fact and of no other and, further, that the fact could not be other than it is" (i.2, 71b9 = text. 7, *Opera omnia,* i, 1552, f.130^{v}); "Demonstrated knowledge must rest on necessary first principles; for the object of scientific knowledge cannot be other than it is" (i.6, 74b5 = text. 44, f.142^{v}). I should argue that Galileo aimed in the end at total certainty, that it was Aristotle and no other who provided him with this ideal of truly scientific certain knowledge, and that he retained this ideal from his earliest to his latest writings, even as he rejected the methods and destroyed the content of Aristotle's physics, and even when he recognized that demonstration truly scientific by Aristotelian criteria eluded his grasp.

We might say that by attempting to prove so much so powerfully Galileo got himself scientifically and personally into a lot of unnecessary trouble. But given his background and education in sixteenth century Italy, to say nothing of his own quite specific intellectual vision, it was very natural for him to see beyond the solutions of particular problems to a general philosophical reform to which they would effectively contribute. In this he was certainly encouraged by early influences to make a characteristic response to the striking variety of current intellectual attitudes and aims, themselves the products of successive European responses to successive recoveries of ancient thought.

Most relevant was the well-known difference between the philosophers on the one hand, and the mathematicians and artists on the other. Both sides had been exposed in different ways to a mathematical rationalism imposed on art and nature through mathematical theories of painting, music and machines, and on philosophy through Neoplatonic visions of a morally normative and therapeutic numerological harmony, and of mathematics as a stage in the education of the mind for theology. Mathematics became an antidote to the threat of scepticism. But the recovery of alternatives to the academic Christian Aristotle, and especially of this new Plato, made much sixteenth-century philosophy notably eclectic, tolerant of opposing systems, seeking concordance between authorities, circling in the habit of scholastic disputation, seeing mathematics as a means of moral education rather than of solving scientific problems. Jacopo Mazzoni (1548–98), friend of Galileo's father and professor of both Aristotelian and Platonic philosophy at Pisa from 1588 to 1597, was the most obvious and intelligent philosophical contemporary giving mainly this meaning to mathematics.

By contrast the artists, engineers and mathematicians concerned with their problems were obliged by their practical crafts to make clear limited decisions. The Florentine ambience provided by Galileo's father, as an eminent practical as well as theoretical musician, and by his friends among artists and mathematicians, was strongly scientific in this sense and unsympathetic towards the more numerological and cosmic aspects of Platonism. Moreover, Vincenzo Galilei (c. 1520–91) in his experimental analysis of the mathematical basis of music looked beyond the Pythagorean proportions, like Aristotle, for some process of physical causation. We could say perhaps that Galileo Galilei tried to carry the decisiveness of the mathematical arts into natural philosophy through the discovery of true processes of physical causation, as distinct from those accepted by conservative contemporary Aristotelians. Out of this, above all under the guidance of Archimedes, came the distinction he made between what he called mathematical "definitions" (e.g. *Discorsi* on two new sciences, 1638: *Opere,* viii, 197 sqq.) and the physical causes which he never ceased to look for. He was to carry the consequent decisions of his natural philosophy into theology. His earliest surviving philosophical writings show however an influence on his intellectual formation that was neither mathematical, nor artistic, nor Platonic but conservatively Aristotelian. To these I must now turn.

During 1969 and 1971 my colleague Adriano Carugo, then working at Oxford and now at the University of Venice, and I solved the main problem of the sources of Galileo's early writings in his own hand, published by Favaro as *Juvenilia (Opere,* i). These comprise two incomplete treatises, each in two parts, on major Aristotelian themes: the *Tractatio prima de mundo* with the *Tractatio de caelo* concerned essentially with questions of cosmology and cos-

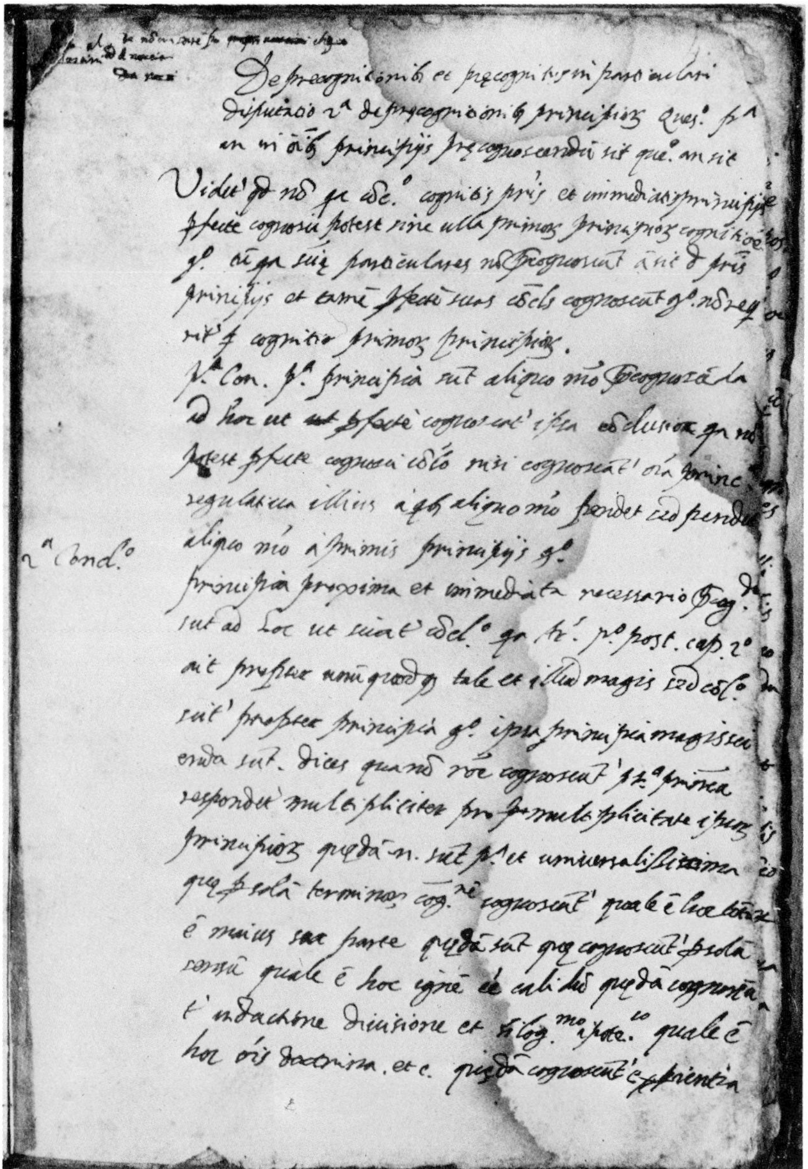

Fig. 1. Beginning of Galileo's autograph *Disputationes de praecognitionibus et de demonstratione (Biblioteca Nazionale Centrale di Firenze, MS Galileiano 27,* f. 4r).

mogony raised for Christian theology by Aristotle's *De caelo;* and the fragmentary *Tractatus de alteratione* with the *Tractatus de elementis* concerned with the theory of elements and qualities put forward by Aristotle in the *Physics* and the *De generatione et corruptione.* We have also studied a third autograph treatise, again incomplete and in two parts, which Galileo left in manuscript but of which Favaro published only a small section, describing it as "some scholastic exercises" (*Opere,* ix, 273). This is the *Disputationes de praecognitionibus et de demonstratione* (Biblioteca Nazionale Centrale di Firenze, *MS Galileiano 27; Fig. 1),* a commentary on Aristotle's *Posterior Analytics* with a detailed analysis of questions of the logic connecting cause with effect, of types of scientific demonstration, and of the relation between mental assent, as in a mathematical proof, and demonstration of actual existence. I shall summarize our conclusions about the sources, dates and nature of these three treatises, and then briefly discuss some of the philosophical views Galileo expressed in them and their relation to those he expressed in later life.

I myself began studying the *Tractatūs de alteratione et de elementis* in 1964 when I was looking for the sources and earlier thinking behind the famous distinction which I discussed in my article, "The primary properties and secondary qualities in Galileo Galilei's natural philosophy", for the *Saggi su Galileo Galilei* and in more detail in the unpublished volume *Galileo's Natural Philosophy* in which Adriano Carugo collaborated, both completed in 1968 (see the Bibliographical Note). By that date I had become interested in a further range of ancient, medieval and more recent sources cited in Galileo's treatise as well as in the *Tractationes de mundo et de caelo* and the *Disputationes,* of which I began to make a preliminary study and got a microfilm in the autumn of 1967. The next stage in this story was that in 1968 Adriano Carugo began to suspect and in 1969 showed conclusively that many of Galileo's citations of ancient and

ex enim [illegible], dicimus ex: astra plura oppositis
motib. se movere ex pluris mult.o corpor. celestiu mult.e
necess.o colligimus cu stelle moveri no possint move
aut in aere, ut quida male censuerunt cu moveant ad
motus motor. orbiu, et q.a no posset assignari ratio cur
contrarijs inter se motib. moveret.; ad 1.m in cõtr.; Remus
cu ipso met: in re met in reb. astronomicis astrono
mos ee consulendos. neq dicas cu eodem ibidem c. 8 orbem
celeste ee propter motu astri, ac proinde ee 8 tantu celo
cu eo. non h. in 9. et 10. celo nullum sit astr: illis in
motus redundat in motu aliquem astror; propterea quod ne
gari posset illud assignari.

Q.o 2.a de ordine orbiu celestiu.

2.a op. — Fuit: quadrigentis annis ante Ptolom: qui ex recetis se
cutus e Nicol: Copn: in op. de revolutione orbiu celestiu
hunc ordinem constituit. ut sol in medio mundi sit collocatus
circa quem orbis [illegible]. circa hunc orbis mag.us terra cu elem.is
et lunam continens circa hunc orbis 8.s deinde celu iou:
postea saturni globus postremo firm.u

f.a op.o confutat. — Haec op.o adversat. oi philosophi: astron: 3.o cont.a et ratiõ terrã in me
dio mundi constituere audent: p.o q.a ex Aris: in 2.o celi, et ex
Ptol: dic: 1.o c.s. si terra no esset in medio mundi ita no fierent
semper eclipses lunae, q.do dua luminaria p diametru oppo
nunt.; sed plerumq contigeret q.do no existit in locis Zodiaci
oppositis; pt. in: pugnat cu expe.a astronomos docentiu tame
semper fieri eclips: q.do luna opponit. soli, alias nunq. cõs.r
q.a si ponat. terra in medio mu di tunc et fiet eclipsis cu oppo
nunt. 2.o luminaria ob terrae interpos.e ita si ponat. extra
mediu no fiet eclipsis q.a nulla erit vi interpos.o terrae. imo
tunc maxe fiet q.do luna maxe distabit a Sole, q.a tunc
habebit minimu luminis. 2.o q.a ex regiom.o i epitome

Fig. 2. Autograph page of Galileo's *Tractatio de caelo* with the earliest reference in his hand of Copernicus's great work: "Nicol. Copn: in op. de revolutione orbinum caelestinum" *(Biblioteca Nazionale Centrale di Firenze, MS Galileiano 46,* f. 22r).

medieval sources in the *Tractatūs de alteratione et de elementis* and the *Tractatio prima de mundo* came from the text-books of two Jesuit professors of philosophy at the Collegio Romano, both Spaniards: Benito Pereira (c. 1535–1610) and Francisco de Toledo, or Toletus (1532–96), who became a Cardinal. These text-books were Pereira's *De communibus omnium rerum naturalium principiis et affectionibus libri quindecim* (published at Rome, 1576; first edition with a different title 1562), and Toletus's commentaries on Aristotle's *Physics* (published at Paris, 1581) and *De generatione et corruptione* (published at Venice, 1579). Carugo showed that Galileo again used Pereira's book as his main source of information for his discussion in *De motu* of the dynamical theories of Philoponus, Hipparchus, Avempace, Averroës, Julius Caesar Scaliger and other ancient, medieval and more recent authors. Then in June 1971 I discovered that important parts of the *Tractatio de caelo,* including the earliest appearance in Galileo's hand of the name of Copernicus *(Fig. 2),* whose location of the Earth in an orbit round the Sun is there rejected, all came from a well-known text-book by another Jesuit professor at the Collegio Romano, *In Sphaeram Ioannis de Sacro Bosco commentarius* (published at Rome, 1581) by the German mathematician Christopher Clavius (1537–1612). So Galileo's basic sources were three prominent contemporary Jesuits of the Collegio Romano.

These identifications required some luck as well as cunning, for although Galileo clearly indicated Pereira as a source, he named Clavius only once and Toletus not at all; but of course they were based essentially on considerable and sometimes tedious reading of sixteenth-century natural philosophy, made in order to explore and understand Galileo's intellectual background and its relevance to his own thought. Sometimes Galileo took from his sources whole passages verbatim, including lists of references, not always copied accurately. Sometimes he went through these to the ancient or medieval originals. But he did not simply copy, but organized and often rearranged the materials for his own sharply independent arguments. I have shown that he used another work by Pereira, a commentary on *Genesis* (first volume, published at Rome, 1589), in the same way for his discussion in his *Lettera a Madama Cristina di Lorena* (1615) of the exegetical rules for relating demonstrated science to the authority of revealed Scripture. In the *Disputationes* he cited some two dozen ancient, medieval and more recent authors, but again he seems to have used intermediate sources, here mainly the Dominican philosopher Thomas de Vio Caietanus's *In . . . libros Posteriorum Analyticorum Aristotelis castigatissima commentaria* (published from 1505 in many editions, including one at Venice in 1565) and the sixteenth-century Averroïst logician Girolamo Balduino's *Quaesita aliquot . . . logica et naturalia* (available in various editions including one published at Venice in 1563 with his commentary on the *Posterior Analytics*). We have a complete transcription made during 1970–71 by Adriano Carugo of the unique manuscript of the *Disputationes (MS Galileiano 27),* and we are publishing with an English translation the parts of this most relevant to scientific thought.

All three treatises comprise closely reasoned arguments, scholastic in form, making often fine distinctions between opposing opinions. Apart from Aristotle, cited continuously, the highest rates of citation are scored by his commentator Averroës, followed by Aquinas and the "Thomistae" (*Opere,* i, 76, 117–118, 144), chiefly Italian and Spanish. This is matched by agreement with Thomist opinions especially on cosmology, for example for the world created being the best possible, for the heavens being probably incorruptible but not necessarily so because no natural power could limit God's absolute freedom, and so on. If we look at Galileo's Jesuit sources themselves, we find an astringently rational view of nature, natural causation and natural philosophy very like so many later expressions of his own. Pereira, for example, argued that the disproof of alchemical gold came not from the theory that alchemists had no access to celestial fire, which he himself thought was the same as terrestrial fire, but from the fact that no one had ever produced it (*De communibus,* viii. 21, pp. 299–300). He was equally sceptical of magic and astrology. Clavius gave a brilliantly lucid exposition of the criteria for deciding whether or not the spheres and epicycles, postulated in astronomical theory to account for the observations, had any real physical existence (*In Sphaeram,* c.4, pp. 434–437). Galileo did not discuss this in the *Tractatio de caelo,* but we may see a kinship between his later position on Copernicus and Clavius's insistence that celestial like terrestrial science must argue from effects to their real physical causes, that it was only the syllogistic form that made the dialectical rule that truth can follow from falsehood seem plausible, that Copernicus himself had postulated his new arrangement of spheres and epicycles not as fictitious but real, and that while he himself was not convinced by Copernicus' arguments he would thank heartily anyone who could produce a better system than any so far produced.

What are these writings? We have derived a possible order and dating from their content and from the paper used. The chronology in the *Tractatio prima de mundo,* deriving from a combination of Biblical and ancient Greek chronology a total of 5,748 years from the creation, with 1584 years from the birth of Christ "down to the present time" (*Opere,* i, 27; *cf.* Favaro's editorial comment on p. 12), might be thought to make this at least its earliest date of composition, even if it was all copied from another source. Since the *Tractatio de caelo* is written on the same kind of paper, watermarked with a faint CT or CL, they seem to belong to the same period. Should both be placed at the end of Galileo's period as a student at Pisa, before his return to Florence in 1585? But, as William Wallace has pointed out to me, Galileo corrected mistakes in writing down this total chronology (*MS Galileiano 46,* f.10^{r}) and when repeating it later wrote it as 6,748 years without correction (f.15^{v}; corrected in *Opere,* i, 36), so it seems to be fragile evidence. Moreover, in the *Tractatio de caelo* he quoted Clavius. He visited Clavius in Rome in 1587 and evidently discussed astronomy, for in a subsequent letter of 8 January 1588 (*Opere,* x, 22–23) he referred to the Jesuit's still unpublished defence of the new Gregorian calendar. In his letter of 15 November 1590 (*Opere,* x, 44–45) to his father from Pisa, a year after he had returned there as lecturer in mathematics, he awaits the arrival from him of "la Sfera", which could have been Clavius's. So perhaps we should date the *Tractationes de mundo et de caelo* from his period either with his father at Florence (when in 1588 he wrote his cosmographical lectures on Dante's *Inferno,* on different paper however) or as a young lecturer at Pisa.

The *Disputationes* is written on paper without watermark. Since here he does not mention Archimedes, explicitly the new enlightenment of his *Theoremata circa centrum gravitatis solidorum* (dated late 1587 or early 1588: see Carugo's edition of the *Discorsi,* 1958, pp. 840–847) and thereafter of the lectures on the *Inferno,* the dialogue and treatise *De motu,* and *La bilancetta* (dated 1586 by Favaro on Vincenzo Viviani's not always reliable testimony, but plausibly later on other evidence to be discussed in our forthcoming book), it seems that the *Disputationes* must probably precede these works. Of these *La bilancetta,* the *Dialogus de motu* and part of the *Tractatus de motu* were also written on similar paper without watermark. He wrote the *Tractatūs de alteratione et de elementis* on the kind of paper, watermarked with a device of a lamb and flag (Fig. 3), which he used also for the *Inferno* and for another part of the *Tractatus de motu.* It has been argued, mainly from the doctrines proposed, that he wrote both the dialogue and the treatise *De motu* after his return to Pisa in 1589. If the paper is a guide to the date of the *Tractatus de elementis,* this would connect the sudden appearance of citations of Galen in this work with the seven volumes of Galen which Galileo said in the same letter of 15 November 1590 that he was expecting from his father with the *Sfera.* Some years after giving up medicine, it was Galen the philosopher whom he cited.

In this letter he told his father that he was "studying and having lessons with Signor Mazzoni, who sends you greetings". Must we then conclude that the *Tractatūs de alteratione et de elementis* was a study of these questions of Aristotelian natural philosophy written by the young lecturer in mathematics under the influence of Mazzoni, side by side with the critique of Aristotle he was developing in *De motu* under the influence of Archimedes and Plato? The targets for criticism are also indicated by Mazzoni: Aristotle's lack of mathematics and his uncritical reliance on the senses. Galileo contrasted both with his own new mathematical method, but neither criticism is incompatible with his making at the same time a serious study of Aristotle's theory of the elements and qualities and its ancient rivals.

In the unpublished volume I have already mentioned, I suggested that Galen's exposition of atomist doctrines in his *De elementis secundum Hippocratem* could have been a source of Galileo's later distinction between primary properties and secondary qualities which he had known from that time. This was also suggested by William Shea in his article "Galileo's atomic hypothesis" (*Ambix,* xvii, 1970, p. 23). Moreover in *De motu* itself Galileo retained scholastic forms of argument alongside the mathematical form learnt from Archimedes, and continued not only citing philosophical commentaries but also using Pereira as an important source of information. Already in *De motu* Galileo used Archimedes and Plato to replace Aristotle's teleological structure of the universe with a structure that was the resultant, still providentially designed, of mechanical forces, and at the same time to begin replacing the whole Greek theory of pairs of contrary qualities with quantitative linear scales of

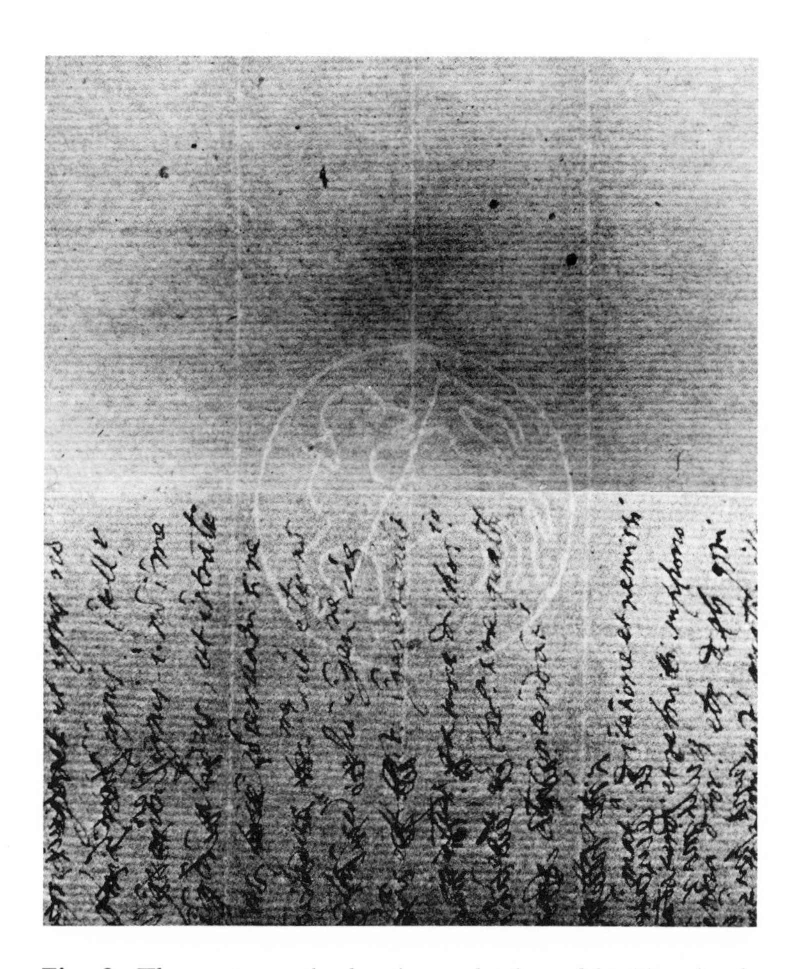

Fig. 3. The watermark showing a backward-looking lamb with flag enclosed in a circle: Briquet no. 48 *(Biblioteca Nazionale Centrale di Firenze, MS Galileiano 46,* ff. 71, 74: the paper is folded and bound across the middle of the circle).

weight, density, heat and so on. The full integration of his new mathematical method with a new theory of matter was something he brought about only much later, precisely through a further critique of Aristotle.

We may then dismiss the hypothesis that Galileo's three earliest treatises were notes he took of philosophical lectures heard as a student at Pisa. The long-standing candidate for the lecturer, Francesco Bonamico, has in any case been shown by Eugenio Garin (*Scienza e vita civile,* 1965, pp. 124–127, 144–145, 165–166) to be impossible, and this was confirmed in 1969 by Adriano Carugo's further comparisons of Bonamico's *De motu* (Florence, 1591) with the *Juvenilia.* Bonamico was no Thomist and he disagreed with Galileo too often. Galileo was to take him on years later in his *Discorso* (1612) on floating bodies, and interestingly was to cite from him the logical rule for discovering the causes of effects through presence or absence, which he used in experiments for that work (*Opere,* iv, 52; *cf.* 19, 22, 27). But that is another question. I do not think it possible to say what Galileo wrote these treatises for, or indeed exactly when he wrote them. Was he lecturing on these subjects and were they his own lectures? Were they simply for his own edification? For that matter why, and indeed over what years, did he write *De motu?*

Before we made the discoveries I have described no one known to us, no one we had been in touch with or whom we knew to be working on Galileo, had identified any of these sources. It seems that we looked back across nearly four and a half centuries to something known before perhaps only to Galileo himself. But someone was bound to identify them fairly soon, and in fact William Shea did independently discover Galileo's use of Clavius a couple of years after me. William Wallace noticed certain similarities with Pereira and Toletus, but saw them only among others through a glass darkly and failed to identify them as sources. Full details of our work will be published in our forthcoming book, but meanwhile we thought it might be useful to make authorized information available. It seems likely that Galileo used other secondary sources not yet identified. The sheer number of references, not just to ancient, medieval and modern philosophers and astronomers but also to points of theology in Scripture, patristic writings and the decisions of Councils of the Church, suggests some common source. Perhaps someone, not me, will look further. Nevertheless these early writings impress by their scholarship. They show Galileo then as indeed he appears in his later writings (despite his biographers) as the highly literate, well-read man of his time and ambience that he was, a match for anyone in learned dialectical debate, and a philosopher who in wanting to show forth the true system of the universe and of knowledge, wanted also the support of the truest ancient model. He famously asked to be entitled "philosopher" as well as "mathematician" to the Grand Duke on his return to Florence in 1610 (*Opere,* x, 353).

The theory of truly scientific demonstration expounded by Aristotle in the *Posterior Analytics* was a model on which everyone in Galileo's time had been educated and which was widely accepted as the ideal goal of knowledge. Galileo's *Disputationes de praecognitionibus et de demonstratione* was his account of that model. It is significant that he should have written it as one of his earliest philosophical essays. Let me conclude by looking briefly at its place in Galileo's thought.

We are caused to have knowledge, Galileo wrote in *De praecognitionibus,* by the first principles we grasp (*disputatio* ii, *quaestio* 3, *MS Gal.* 27, f.5^{v}). We may know these in various ways: the most universal only through knowledge of terms, as that the whole is greater than its part; others only through the senses, as that fire is hot; others through various forms of inductive or hypothetical argument; others through experience, as in medicine; others only through habit, as those of moral science which we cannot understand unless we practise them (ii.1, f.4[r]). But whereas in nature an effect must necessarily follow from its sufficient cause, man is free and cannot without his assent be made to have knowledge (iv.2, f.12[v]).

This leads to a discussion in the *Tractatio de demonstratione (disputatio* i, *quaestio* i, f.13[r]) of Aristotle's criteria for the first principles of truly demonstrated knowledge: these must be true, primary and immediate in not being themselves demonstrated from any prior principles, and related to their conclusions as cause to effect (*Post. Anal.* i.2). Galileo argued that only true propositions can actually be known, because true knowledge of things is had through the causes by which they exist. Demonstrations of true conclusions from false premisses can only be *per accidens,* not *per se,* and we cannot actually know such things as the void and the infinite for they are nothing. The proper object of true knowledge is *ens reale,* real being, not just *ens rationis* (ii.1, ff.17[v]-18[r]). He went on to analyze at length Aristotle's criterion that truly scientific demonstration must proceed from true causes, though we have first to discover these from our more immediate knowledge, for example through the senses. The premisses of mathematics cause knowledge and are as immediately knowable to us as their conclusions, but mathematical entities do not exist (ii.6, f.22[rv]). The sciences subordinate to mathematics (as astronomy, music etc.) do not have truly scientific demonstrations because they must proceed *ex suppositione* from principles assumed from the superior science (ii.4, f.20[v]). We may give our certain assent with evidence as to knowledge through the senses, or without evidence as in our faith, but we come to rest most agreeably in knowing a conclusion because it follows from true premisses (ii.6, ff.22[v]-23[r]).

He concluded with a discussion of the recognized kinds of demonstration: *ostensiva, ad impossibile, quia, propter quid, potissima* (iii.1-2, ff.29[r]-30[v]; *cf.* i.1, f.13[r]). Here, as elsewhere, he seems to be using Proclus' commentary on Euclid, as well as Averroës and other authors whom he named, but he took an independent line. Demonstration *ad impossibile* is not truly scientific because it proceeds by raising questions from false premisses in order to find the true ones (f.29^{r}). Truly scientific demonstration could be reduced to two kinds, *demonstratio quia* which demonstrates the existence of an effect and from that *a posteriori* its cause, and *demonstratio propter quid* which demonstrates both the cause and hence the existence of the effect (f.30^{rv}). That *demonstratio quia* is truly scientific is proved on the authority of Aristotle and all commentators, and because like *demonstratio propter quid* it proceeds from true and necessary premisses to true and necessary conclusions, and so generates knowledge and not probable opinion (f.30^{r}). That an attribute is connected with a subject we know from experience; that the connection is naturally necessary when it always occurs we know by the light of our intellect, for otherwise nature would have been improvident; it can be truly demonstrated by intrinsic, extrinsic or other kinds of cause (f.30^{v}). This seems to be the origin of Galileo's later designation of demonstration both from observation and from theory as "necessary demonstration".

The scientific argument, he went on, especially in the physical sciences where we began by not knowing the physical causes, alternated in a "demonstrative regress" (iii.3, f.31^{rv}) in both directions, from effect to cause and vice versa. In mathematics the regress is little needed because premisses are as immediately known as their conclusions. In any case it is not circular because, starting from an effect which one knows better than its reason, it demonstrates the reason for that effect. The complete true cause and the effect entail each other reciprocally and uniquely (f.31^{v}).

Parts of the *Disputationes* (despite its containing no precisely scientific illustrations of the logic) resonate with many of Galileo's well-known later practices and sentences. This is not the occasion to discuss the organization of his experimental argument, for example in *De motu* and in the *Discorso* (1612) on floating bodies, on the logic of *la progressione demonstrativa,* the *methodo resolutiva,* and the *reductio ad impossibile* or *ad contradictionem* (*Opere,* i, 260–265, 284–285, 318; iv, 19, 22, 27, 67). But it is relevant to note that he continued to carry on about "true and necessary demonstrations" and "the necessary constitution of nature" (as he put it in *Le mecaniche,* 1593; *Opere,* ii, 155, 189), and "true demonstrations" from "the true, intrinsic and total cause" (*Discorso,* 1612; *Opere,* iv, 67), from his earliest writings and throughout the telescopic, mechanical and Copernican debates of 1610–16 and down to the *Dialogo* (1632). The great attraction for him of his argument, first put forward in 1616, from the tides to the Earth's motions seems to have been that here he had a truly scientific demonstration by Aristotle's criteria: this cause must produce those effects, and those effects must entail this cause and no other (*Opere,* v, 377–381, 393; vii, 443, 470–472). Galileo hedged by claiming this as perhaps only the most probable cause advanced so far, but he exposed himself of course to a double accusation: that he was committing the logical fallacy of affirming the consequent, for phenomena could not uniquely determine their causes; and that he was claiming to demonstrate something necessary not just about the world that existed but also about its omnipotent Creator (*cf.* Antonio Rocco in 1633 on the *Dialogo: Opere,* vii, 628–629, 699–700).

Galileo's necessity surely belonged to a conception inherited from Greek philosophy, that of the possibility of a completed and bounded knowledge of all that does and can exist. God's omnipotence made this existentially untenable, and this Galileo was to be careful to accept, by distinguishing his arguments about the world God had in fact created from any suggestion that God could be bound by any natural necessity (*cf. Dialogo: Opere,* vii, 128–131, 488–489; *Lettera a Madama Cristina di Lorena: Opere,* v, 316–321). In his scientific practice, the open-ended character of mathematics and experiment and of the Archimedean argument *ex suppositione* (as in his letter of 7 January 1639 to Baliani: *Opere,* xviii, 12-13, aptly quoted by Stillman Drake), his appreciation of the complexity of natural causes themselves in such phenomena as light and heat, above all his use of range of confirmation as the test of a theory, notably of the new cosmology, effectively killed the scientific ideal of necessary truth imposed by Aristotle's logic. What are we to make then of Galileo's apparent blindness to this in expressions of continuing hope? Perhaps just words. But it seems to me that we have here in the slow general understanding of the difference that mathematical thinking made to traditional logic and to scientific explanation, found after all in sixteenth-century attempts to put Euclid into syllogisms, a phenomenon in European intellectual history, in European scientific methods mediated through cultural habits and inherited preconceptions, that greatly merits attention.

Mathematics and Galileo's Inclined Plane Experiments

PIERRE COSTABEL

Inasmuch as I believe that Professor Drake has rightly called into question a too systematic use of the notion of mental experiment, allow me to open my remarks by saying a few words about the systematic use of manuscripts.

For many years I poured over the manuscripts of Malebranche, and it was only after discovering my early mistakes and finding evidence of the foolishness of some of Malebranche's contemporary interpreters that I came to realize what I was really doing: prying into the secrets of the dead. Since then, I have come to believe that the study of manuscripts, which is so important for our understanding of history, cannot be undertaken with nothing more than analytical ability and a thorough grounding in the technical skills required for such an operation. Something else is needed: a deep respect for those who have come before us. Even when we are dealing, as in the present case, with purely scientific documents which do not call for moral evaluation, reverence for the author has the advantage of reminding us of the need for rigour, caution and objectivity.

The survival and transmission of rough notes, the jottings of a mind at work, is largely a matter of luck. By their very nature, these notes were meant to be ephemeral and when they reach us they can only speak inarticulately. As partial records of a discussion that an author carried on with himself, they are by themselves broken chains of thought. *Rigour* is necessary to acknowledge their fragmentary nature, *caution* to resist the desire to fill the gaps with logical connections, and *objectivity* to limit our pronouncements to what the documents actually disclose.

With this in mind, let us turn to folio 116*v* of volume 77 of the Galilean manuscripts which has played a major role in the basic orientation of Professor Drake's research. The date 1608–1609 that he suggests for its composition is a result of his careful examination of the watermarks and the paper, and I am happy to rely on his expertise. You will recollect that the main figure on this folio shows parabolic paths that appear to represent horizontal throws. Their distances, on a horizontal line whose position is well defined from the point where the motion originates, are expressed in whole numbers and are accompanied by a few words (see Fig. 4 in Drake's article, p. 140. There is no doubt here: the author has clearly indicated that these numbers are different from those he expected and which he also transcribed on the folio. Here and there on the same page, numerical calculations reveal the formula that was used.

What is clear from this folio is the formula used to calculate the results on the one hand, and the comparison of these results with others obtained *by some other means* on the other.

The formula states that the squares of the horizontal distances are proportional to the numbers on the vertical line. These numbers would seem to indicate the height from which the body was dropped and it is all too tempting to conclude that we are presented with an analysis of the conversion of vertical motion into horizontal flight. The ease with which we take this step depends, of course, on what we know of Galileo's published works, notably of his discussion of the Platonic cosmogony in the *Dialogue* and the *Discourses* which is based on such a conversion. If we follow this train of thought, we shall be led to add that when Galileo jotted down these notes, he visualized the speed of a freely falling body both as proportional to the square root of the time of descent and as converted into uniform horizontal motion.

Note that we thereby introduce a number of relations that follow logically from our more advanced physical viewpoint but that the documents themselves do not necessarily bear out. In other words, was Galileo aware of this network of physical relations?

But this is not all. When we compare the results calculated by Galileo with the numbers I have referred to as "obtained by some other means", we are prone to consider the latter set as the outcome of an experiment. But what device, i.e. what kind of inclined plane did Galileo use? And if Galileo did roll a ball on an inclined plane, how are we to account for the discrepancy between the first set of (theoretical?) figures and the second set of (experimental?) figures? Naturally, we cannot neglect the reduction of the rate of acceleration for rolling (without sliding) that our more sophisticated modern theory enables us to work out but that Galileo was not in a position to measure. This slowing down of the acceleration explains why experimental results are always *inferior* to theoretical expectations. But here is the snag: the results expected — namely those that Galileo calculated — fall short of the numbers we tend to interpret as experimental results. I believe that this is the crux of the matter, and that the first problem to solve is the precise significance of the set of numbers: 800, 1131, 1306, 1329, 1460, such that:

$$\frac{(800)^2}{300} = \frac{(1131)^2}{600} = \frac{(1306)^2}{800} = \frac{(1329)^2}{828} = \frac{(1460)^2}{1000}$$

Without going into the details, which are unnecessary for our purpose, we can say that this set of proportions corresponds to what we know of the relation of a uniformly accelerated motion to the time: the denominators stand for the distances travelled and the square roots of the numerators for the results, during the same period of time, of uniform motions equivalent to each of the distances.

It follows that this series of numbers is independent of the inclination of the plane on which the ball is made to roll. Whether 300, 600, 800, 828 or 1000 stands for the height along the vertical or for the length of the incline, we obtain the same series: 800, 1131, 1306, 1329, 1460 for the distances of the corresponding uniform motions during the same period of time.

Note that we make things difficult for ourselves precisely by considering the results as experimental. The difficulty consists in ascribing a definite value to "the same period of time" that guarantees identity of results whether the numbers along the vertical axis be taken for the distances along the incline or along the upright side of the inclined plane. The words on the figure: "828 height of the table" would seem to indicate that projection originated from a table and that the horizontal distances are calculated for a period that is identical to the time a body takes to fall through 828 points.

But if the time is determined in this manner, are the calculations still correct? The answer is, no.

We can see this by considering the first distance of 800. I spare you the elementary calculations that establish that half this distance should be the mean proportional between 828 and the height of the drop prior to the conversion of the motion into horizontal projection. Namely $(400)^2 = 828$ h, where $h = 300$ or $300 \sin \theta$ if we assume that 300 is the distance on the plane raised at an angle θ. Now a solution is only possible in the second case with:

$$\sin \theta = \frac{1600}{2484} = \frac{2 \times 800}{3 \times 828}$$

But this value has too little meaning in a trigonometrical context for us to assume that it could have been used by Galileo either in the elaboration of his theory or in making his experiments.

From the standpoint of theoretical simplicity and experimental ease, we may note that if the series 300, 600,

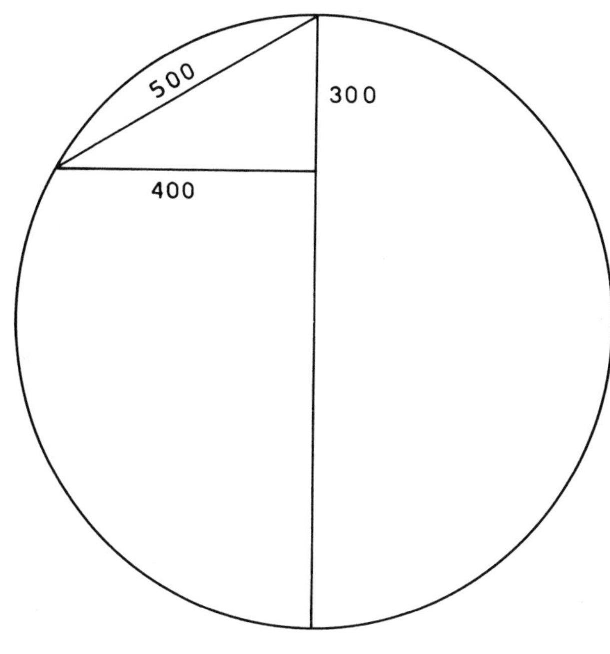

Fig. 1.

800, etc. stands for the heights of the initial vertical drop (and this seems to be the most obvious meaning of the figure), and if we assume an inclined plane with a slope of ¾ (the simplest to devise), we obtain very plausible results by replacing 828 with 833 which is the diameter of the circle in Fig. 1. This diagram illustrates what was familiar to Galileo: the time of free fall along the vertical of 833 points is equal to that of the descent along the inclined plane of 500 points. Now this descent along the incline corresponds to a horizontal distance of 400 points. It is tempting to suggest that this horizontal distance (multiplied by 2) determines the distance of the subsequent horizontal projection during an equal period of time! This is wrong, of course, since the horizontal projection is $2 \times 500 = 1000$, but we see how easily such an error could have been made.

It would seem, therefore, that when we interpret folio 116*v* as representing the transformation of motion along an inclined plane into horizontal motion, we do so at the cost of ascribing two errors to Galileo. The first consists in taking 828 instead of 833, the second in equating descent along the incline with its projection on a horizontal line.

This consideration has the advantage, however, of allowing us to modify the numerical results that Galileo should have expected by replacing 800 with 1000 as the first term of the series. This yields relatively higher values for the set of numbers *obtained by some other means* and which, as we have seen, appear to have an experimental origin.

The justification of this lies in the significance of the discrepancies between the two sets of numbers. Let us consider the second set which lacks a first term and whose correspondence to the heights along the vertical line is unambiguous:

	1172	1328	1340	1500
h =	600	800	828	1000

Are these numbers the outcome of experiments as Prof. Drake claims in the light of the data on folio 175*v*? I feel uncomfortable about this interpretation for nowhere, in the relevant Galilean manuscripts, do we find measurements in the usual units and their conversion into *points.*

I agree that if we assume, with Professor Drake, that the point is the Florentine unit of 17/18 mm., then the numbers correspond to lengths that are reasonably close to those that we obtain if we perform the experiment in our laboratories. But the unit, for Galileo as well as for his modern interpreters, can remain unspecified. In other words, what we find on this page of folio 116*v* are nothing else but proportions. We cannot, a priori, discard the hypothesis that they may be the outcome of abstract mathematical reasoning.

What folio 116*v* provides is strictly the following data: a figure and two series of numbers which are related to heights. Hence:

(I)	1500	1340	1328	1172	
(II)	1460	1330	1306	1131	800
h	1000	828	800	600	300

The path of (II) is parabolic and the first value calculated in this series is 1131. The series runs, therefore, from right (800) to left (1460). It is striking that the roughly parabolic line that leads to 800 is dotted and not continuous, as if the author wished to indicate that it does not have the same status as the others (see Fig. 1 in Drake's article).

The numbers in both series are plotted on the graph in Fig. 2. The terms of series (II) lie along a parabola P, but the terms of series (I) do not. This calls for two comments. First, as we have already mentioned, the discrepancies between the numbers of (II) (anticipated results) and the numbers of (I) are not of a kind that we would normally expect in experimental situations. They are not *lower* but *higher* than the predicted figures. Second, it is impossible to con-

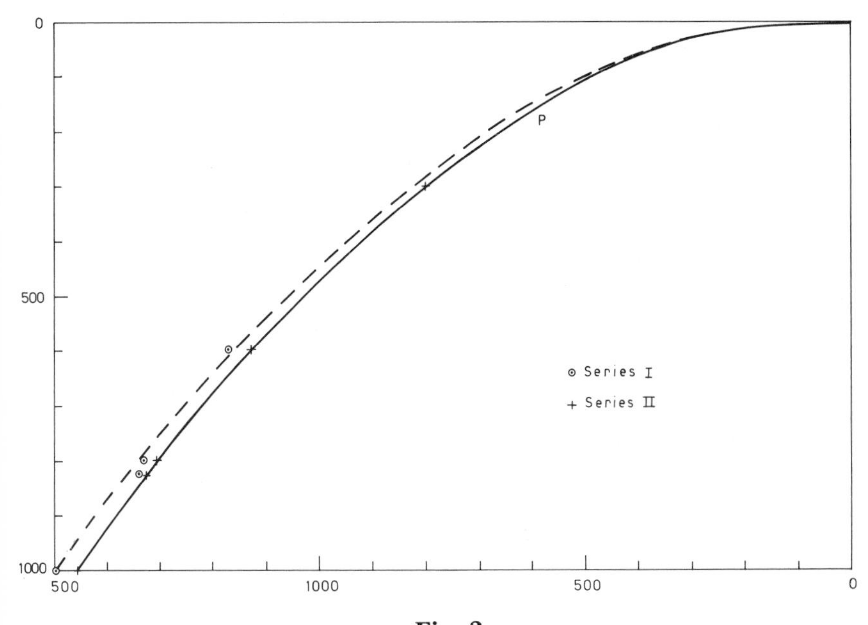

Fig. 2.

nect the numbers of (I) with a smooth curve having the same shape as the parabolic path P of (II). Now if (I) were to be understood as the outcome of an actual experiment, we would have a very curious situation indeed, namely we would have to surmise errors in measurements (by excess for h = 600 and h = 1000, and by default for h = 800 and h = 828) that are relatively large.

These two considerations lead me to ask whether the series (I) does not result, at least in part, from mathematical speculation.

1172 is the only number accompanied by a complete sentence: "1172, should be, to correspond with the first, 1131". Now since 1131 is the first value calculated in series (II), it is clear that "the first" refers to 800 and that 1172 is not obtained by the same method as in series (II) where the order runs from right to left. Otherwise Galileo would *already know* that 1172 does not correspond to 800 and he would not raise the question. A comparison of the terms of (I) with those of (II) confirm that (I) is to be read in the inverse order from (II). It would seem that Galileo proceeded from left (1500) to right and having reached 117 asked himself how to arrive at 800, and then went through the series (II) from right to left:

$$\text{(I)} \quad 1500 \rightarrow 1172$$

$$\text{(II)} \quad 1460 \leftarrow 1131 \leftarrow 800$$

On what grounds can we maintain that 1500 and 800 are related? Because 1500 − 800 = 700 and 700 = 1000 −300, the difference of the heights *h*. Inasmuch as the numbers of the series seem to indicate speeds acquired in falling from heights *h,* relating 1500 with 800 consists in checking whether a linear function obtains between h = 300 and h = 1000.

This interpretation tallies with Galileo's letter to Sarpi of 1604 in which he writes that he believes he has found a way of relating the times-squared law with a speed "that increases with that proportion with which it departs from the beginning of its motion". Although we should not be overhasty in concluding that this means that the speed is proportional to the distance, nevertheless, there can be no doubt that a linear function of space is the simplest interpretation that can be given to Galileo's own words. If we recollect that mathematicians of the period did not think in terms of functions but in terms of tables of numerical values, we can see why a table in which the difference between the extreme values (1500 and 800) is equal to the difference in the heights of free fall (1000 and 300) provides a normal test of Galileo's claim in his letter to Sarpi. In this light, it is revealing that the number of series (I) are located in the region where the curvature of the parabola P is minimal and where linear approximations are at their best (see Fig. 2).

Finally, the numbers of series (I) and their corresponding heights display curious relations. For instance, $1172 = 828\sqrt{2}$, $828 = 2 \times 1000 \times (\sqrt{2} - 1)$, and $1500 - 1328 = 172 = 1000\,(\sqrt{2} - 1)^2$.

From these relations, I have not yet been able to ascertain how series (I) was actually formed, but their very existence makes it unlikely that the series is merely the result of experimental measurements. The number 828 plays a special role and even if we admit that the phrase "*altezza della tavola*" on folio 116*v* suggests a physical table (namely if we do not consider it as a difference in a *table* of numerical values), it is hardly plausible that it should have been arbitrarily chosen.

It is difficult, therefore, to say anything decisive about folio 116*v*. I do not wish to exclude that Galileo made experiments, but I do not believe that mere empirical measurements can account for the message contained on this page. In other words, I suspect that the relation between mathematical speculation and experimental testing is more complex than Professor Drake assumes. Until further evidence is forthcoming, I shall continue to believe that the Galilean manuscripts on motion bear witness to a *special kind* of experiment: a quest for the underlying unity between different ways of mathematically describing the phenomenon of free fall. This would simply be the implementation of the research program implicit in the letter to Sarpi. It is a matter of *experiment* not of mere speculation, for struggling with mathematical reality is not the same as inventing abstract theories.

A question of method is at issue here. In my critical edition of Marin Mersenne's *Les nouvelles pensées de Galilée* (Paris: Vrin, 1972), I had occasion to show in the notes how Mersenne was unable to free himself from a *physical* consideration of free fall, from an analysis of motion that can be actually experimented and translated into mathematical relations. He failed to see that Galileo's motion *naturaliter accelertus* was first and foremost a motion defined according to a *coherent* mathematical network and that experimental testing came only later. He did not grasp that *naturaliter* had a new meaning.

This is an important aspect of the Scientific Revolution and it is at the heart of the debate on the principles of mathematics and the foundations of Euclidean geometry that run from Ramus to the *Logique* of Port-Royal. A close examination of the real process whereby a mathematician arrives at his definitions and his rigorous structures revealed that it cannot be reduced to deduction. "*Natural*" qualifies this process, and, in this sense, although coextensive with "logical", it has a different import. This seems to me to be one of the major methodological advances made in the seventeenth century. The question whether the new science appeals to experiment to test the agreement between the natural order of things and the order elaborated by the mind of the interpreter presupposes an answer to this other question: Is rationality, basically, a matter of internal coherence? And is this coherence both the hallmark of the new "natural" science and the condition of its success?

The Role of Alchemy in Newton's Career

RICHARD S. WESTFALL

For well over two centuries now, the word has been out that Isaac Newton left behind a large collection of alchemical papers. The Rev. William Law asserted as much on the authority of Humphrey Newton not long after Isaac Newton's death; and though, in his anxiety to enlist Newton in the cause of Jacob Boehme, he did minor violence to the details, his report of extensive notes from alchemical writers was correct. Others at the time knew so as well. The bold judgment, "Not fit to be printed" scrawled on the wrappers by Thomas Pellet, who examined Newton's papers for the family after his death, testifies that a contemporary of William Law both knew of the papers and judged them in more conventional eighteenth century terms. So also did Samuel Horsely half a century later by silently omitting them from the *Opera omnia.* What was unfit to be printed in the eighteenth century was nothing short of scandalous in the nineteenth, and only a Scotch Presbyterian conscience could compel David Brewster to confess that his hero had not only copied "the most contemptible alchemical poetry" but had also annotated "the obvious production of a fool and a knave." He found what excuse he could for the inexcusable in the "mental epidemics" of the age, and he scratched up one crumb of cold comfort from the thought that Leibniz also had dabbled in the art.[1] Much has changed in the century since Brewster wrote. Some scholars at least no longer feel compelled to apologize for Newton's alchemical activity. Quite the contrary, the mental epidemic of yore is raging again. Not only has Newton's alchemy become a respectable enterprise, but interest in it has reached the proportion that Kings College, which holds the largest collection of the papers, now invites zealots to study microfilm copies lest the originals be destroyed by sheer wear and tear.[2]

The spate of activity has served thus far more to bring questions into focus than to answer them. The present climate of opinion may leave us more open to the suggestion that Newton practiced alchemy, but the climate of opinion has to do with us, not with Newton. It can only help us to consider the question of Newton's attitude towards alchemy; it cannot answer it. And certainly it cannot answer the larger question which lies behind that one—whatever Newton's attitude towards alchemy, what role did the attention that he manifestly devoted to it play in his scientific career? For one who has immersed himself in the papers for an extended period, even to phrase the question explicitly is almost to despair of answering it. Only those who have not exposed themselves to that dark and turbulent sea, where no familiar shore offers a charted bearing and waves of arcane imagery surge endlessly on every hand, will find the suggestion of despair extravagant. With the despair, however, goes the conviction that the sea must be crossed—that is, if I may draw back from the alchemical imperative to generate ever more imagery, the papers are a basic datum that have to be explored. By themselves, no doubt, they cannot answer the questions of Newton and alchemy conclusively. If they are used with care, however, if they are taken as a whole and not vandalized by seizing and displaying single papers, which are thereby emasculated in their isolation from the total context, they can offer extensive evidence by which at least to clarify, if not conclusively to settle, the questions raised.

When I speak of taking the papers as a whole, I mean several things. I mean watching for traits that are common to the whole corpus. I mean looking for connections that may exist between individual papers. I mean above all recognizing that the alchemical papers have a chronology. As they were not produced all at one time, so also they do not testify to opinions unconditionally valid for every period of Newton's career. Since Newton's scientific thought underwent remarkable development during his career, the chronological pattern of the alchemical papers offers a device by which to relate his alchemical studies to the rest of his intellectual activities. It offers also a device by which to relate them to another set of papers that are explicitly dated, Newton's record of his own chemical experimentation. I propose to explore the relations among these three elements—the alchemical papers, the records of chemical experimentation, and the familiar events of Newton's scientific career.

One of Newton's early notebooks, *Add. MS. 3975* in the Portsmouth collection, suggests a pattern for Newton's early interest in alchemy. Originally, the notebook appears to have been a continuation of the section entitled *Questiones quaedam Philosophiae* in an earlier notebook (*Add. MS. 3996*), a section which records Newton's introduction to the mechanical philosophy of nature while he was still an undergraduate. In the new notebook, Newton entered a number of headings reminiscent of the *Quaestiones*—"Of Colours," "Of Cold, & Heat," "Rarity, Density, Elasticity, Compression, &c."—and under the headings, in the hand of the mid 60's, he entered notes from his reading in the mechanical philosophers.[3] Although Robert Boyle was easily the major source he called upon, the first notes concerned themselves with issues in the mechanical philosophy; they cannot be called chemical. They soon became chemical, however, to the extent that I think of the notebook as Newton's chemical notebook.[4] In keeping with the new chemical tone, he entered a new set of headings—such as "The medicall virtues of Saline & other Praeparations," and "Of other Animall & Vegetable Substances"—and the notes under them, some of which at least date from as late as the mid 70's, are entirely chemical in nature.[5] As the earlier notes changed from mechanical philosophy to chemistry, so the chemical notes changed from chemistry to alchemy. Notes from Boyle gave way to notes from Starkey's *Pyrotechny Asserted,* made in the mid 70's.[6] A new and final set of headings recorded this change—for example, "Gross Ingredients," "First Preparation," "Of ye work wth common ☉."[7] In a hand that belongs to the late 70's, he entered a small number of notes from Eirenaeus Philalethes, *Ripley Reviv'd* (1678) and John de Monte-Snyders, *De pharmaco catholico* (1666). Although other papers and other evidence indicate clearly that the transition from chemistry to alchemy took place in the late 60's, they confirm the pattern of notebook *3975.* In so far as a distinction between chemistry and alchemy can validly be drawn in the 17th century, Newton appears to have introduced himself first into chemistry and to have proceeded thence to the study of alchemy. In the hand of the mid 60's, he composed a glossary of chemical terms—"Acid salts spirits & juices," "Aqua fortis," "Cementation," "Lutum," and the like. The entries under the headings and the recipes for making basic chemicals that he included at the end, breathe the air of sober chemistry. If he included such headings as "Alkahest" and "Projection," he did not enter anything under them.[8] By the late 60's, Newton was reading Basil Valentine, Sendivogius, Philalethes, and Michael Maier.[9] In 1669, he purchased the six volume *Theatrum chemicum,* from which he took voluminous notes in the years ahead.[10] Let me ignore the interpretation of these notes for the moment, and say merely that by the end of the 1660's Newton was a familiar traveller on the by-ways of alchemy.

There is one other suggestion in notebook *3975.* Newton recorded his own chemical experimentation in it. In 1669, he purchased chemicals and two furnaces,[11] and the notebook includes the results of early experiments which appear, by the hand, to date from that time. Later he entered other experiments, frequently dated, on an expanse of empty pages between the headings mentioned above.[12] I shall return to Newton's experimentation later. For the moment let me merely indicate his inclusion of it among notes which were (in part) frankly alchemical, thereby perhaps implying a connection between the two.

Meanwhile Newton's introduction to the art involved a dimension beyond the intellectual. Among his papers is a collection of alchemical manuscripts, in three different hands, mostly of tracts which have never been published. Since Newton corrected a couple of the poems in the collection against Ashmole, where these specific ones were published, numbered quite a few of the recipes, and copied some of the tracts, we can be sure that he studied the collection with care. Since he also utilized a blank space to write a paragraph (utterly lacking in alchemical content) in the unmistakable hand of the mid 60's, we know that he received it early.[13] Where did it come from? Nearly ten years later, his copy of another unpublished manuscript mentioned that "Mr. F" (perhaps, in the Cambridge setting where "Mr." had its own meaning, Magister F) had given it to him in 1675.[14] There lived in Kings College at this time one Ezekiel Foxcroft, an initiate of the alchemical circle that had gathered originally around Samuel Hartlib. Newton later studied Foxcroft's translation of the Rosicrucian *Chymical Wedding,* which he referred to "Mr. F." Whether or not Ezekiel Foxcroft was Newton's initial contact, a considerable number of manuscript treatises that were not available in published form among his papers testify that he was in touch with alchemical circles over a period of at least thirty years.

The treatises of Philalethes circulated initially among Hartlib's group before they were published. Newton had access to some of them in the late 60's, roughly ten years before their publication.[15] He was enabled to make copies of a treatise by Edward Generosus, a Latin version of the letter of Frederick, Duke of Schleswig-Holstein, a treatise named *Manna,* de Monte-Snyders' *Metamorphosis of the*

Planets, a treatise of Jodocus a Rehe followed by letters of A. C. Faber and John Twysden and notes which appear to record Newton's conversations with the two, and an unpublished treatise of William Yworth.[16] Since a number of these tracts were ultimately published by William Cooper, who was a focus of alchemical activity in London, a presumption exists that Newton had connections of some sort with the group. In March, 1683, he received a letter written from London by one Fran. Meheux, which concerned itself solely with the art, and more than a decade later an alchemist from London knew of Newton's existence well enough that he could seek him out to discuss the work.[17]

The student of Newton's career cannot fail to note his intellectual isolation during virtually his entire stay in Cambridge, isolation broken only in his final academic years after the publication of the *Principia.* The catastrophic decline of the university after the Restoration left it an intellectual wasteland as the Cambridge Platonists died off, and even while they survived, Cambridge wholly lacked any specifically scientific community into which Newton might fit. The alchemical papers raise the possibility that his isolation was in fact mitigated by an unexpected community, the clandestine fraternity of alchemists. Indeed, I cannot resist the further speculation that one (though only one) factor in the ultimate decline of his hermetic activity was his success in a more orthodox scientific community following the move to London.

Although Newton was introduced to the art by the late 60's, only a small fraction of his alchemical papers appear to date from this early period. A larger fraction stem from the years between 1675 and 1687. Not only did Newton acquire unpublished manuscripts, such as de Monte-Snyders' *Metamorphosis,* and read more widely in the published literature of alchemy, but he also stepped up the pace of his own experimentation. His dated experimental notes began about 1678. Perhaps the date recorded his move into the chambers beside the great gate where Humphrey Newton tended the furnaces. Although Newton had found a place to experiment before that time, the garden attached to his final chambers made a real laboratory possible. The garden also made the chambers a valuable commodity within the college so that considerable seniority, which alone determined the assignment of chambers, was necessary to claim such a prize. By 1678 Newton had moved several rungs up Trinity's ladder of seniority; maybe he was able to appropriate the chambers when Francis Bridge either left or died that year. At any rate he was living there by the early 80's, and wherever the dated experiments took place, they began about 1678 and continued until 1696.

The latter date may be the most significant of all, especially in the context of the papers as a whole. Although one can trace a rising curve of sheer quantity in the period before the *Principia,* half or more of the papers, that is, more than 600,000 words devoted to alchemy, date from the early 90's, immediately following the *Principia.*[18] This fact strikes me as the most significant conclusion to emerge from a study of the papers in chronological order. Newton's interest in the art was neither a youthful frolic nor an aberration of senility. It fell squarely in the middle of his scientific career, spanning the time of most of the achievement on which his reputation rests. Indeed, alchemy appears to have been his most enduring passion. Whereas other studies could rivet his attention only briefly, alchemy held it without a major interruption for nearly thirty years. After a year, 1664-65, devoted overwhelmingly to mathematics, when he brought forth the seminal ideas of the calculus, Newton began to lose interest in mathematics, and from that time on increasingly required external stimulation to focus his attention on it.[19] Optics concerned him very briefly, only in the late 60's and early 70's, and he never again returned to it seriously. He devoted himself to mechanics and dynamics only for two limited periods, once in the 60's and then in the two and a half years that produced the *Principia*. Meanwhile his devotion to the art flowed on unchecked. The inventory of his library after his death revealed that more than one tenth of its books were alchemical. He was absorbed enough in experimenting in the early 80's that he lost track of the days of the week; his own dates frequently do not correspond with the calendar.[20] Although we recall Humphrey Newton as the amanuensis who wrote out the printer's copy of the *Principia,* his own recollection of his stay in Cambridge focused on the fires and furnaces, and Newton's records reveal that in the spring of 1686, before the final copy of the *Principia* had been completed, he interrupted its composition to work at the furnaces. Immediately after its completion, when he was in London for the hearing before the Ecclesiastical Commission, he visited a Mr. Stonestreet to purchase chemicals.[21] Most important of all, in the early 90's, less than five years after the *Principia,* he engaged himself in a monumental study of the entire alchemical tradition, a study from which a good half of the papers derive. One cannot avoid the question: have we perhaps mistaken the thrust of Newton's career? To us, the *Principia* inevitably appears as its climax. In Newton's perspective, it may have seemed more like an interruption of his primary labor.

With the chronological pattern of the papers in mind, let us turn to examine their content. From what I have said already, it should be clear that no single descriptive phrase can characterize them. Some are mere copies of treatises which Newton presumably transcribed from other manuscripts because they were not available in published form. The fair copy of John de Monte-Snyders' *Metamorphosis of the Planets,* which was never published in English, and in German only in 1700, is one example among a number.[22] One manuscript appears to be Newton's own Latin translation of Didier's *Six Keys,* which was only available in French.[23] The bulk have usually been described as reading notes. So indeed many of them were, but reading notes of a sort that frequently transcended their own nature to become systematic collations and commentaries. Newton's need to organize information nowhere displayed itself more fully than in his alchemical papers. I have mentioned his early glossary of chemical terms and the three sets of headings in notebook *3975*. Sometime near 1670 he laid out a set of twelve headings—concluding with "Multiplicatio"—in another notebook.[24] Although he did not enter many notes in this one, its effort to organize the product of his reading systematically under categories was the forerunner of later such endeavors which culminated in the incredible *Index chemicus* of the 1690's.

Meanwhile analogous features permeate most of the papers. In what were perhaps his earliest notes on Sendivogius' *Novum lumen,* he inserted the pointing hand familiar to every student of Newton's papers beside a particular recipe. The recipe appears verbatim in a later set of notes on Sendivogius.[25] In one of his papers, under the heading *"Materiae Mineralis praeparatio prima & conversio in aquam,"* he was led by a comparison of what Arnold, Ferrar the Monk, Mundanus, Maier, and Mynsicht had said on the subject to the writings of Faber (i.e., Pierre Jean Fabre). From the *Palladium spagyricum* he cited four different pages, widely spaced, incidentally, and not in numerical order. Fabre published *Palladium* in 1624, Newton stated. In 1634 he published *Hercules Piochymicus* in which he had more to say on the subject; seven page citations followed. Further, in *Hydrographum spagyricum* (1639), he said still more, and Newton cited five pages from that work, followed by a larger number of citations from Fabre's *Propugnaculum alchymiae,* to which he devoted 6 manuscript pages, and the *Panchymici,* to which he devoted two. He concluded with a paragraph from Basil Valentine.[26] Reading notes these pages undoubtedly were, but notes which both systematically expounded what Fabre had to say on the metallic spirit, and also related Fabre's views to those of six other alchemists.

Newton was convinced that all the true alchemists were expounding one and the same work behind the obscenely luxuriant growth of arcane imagery with which they disguised it. Part of his problem then was to equate the various symbols—in much the same fashion as he later tried to decipher the symbolism of the Biblical prophecies.[27] The papers abound with passages in which he simply listed alternative names for one and the same substance.[28]

> Now this green earth [he stated as he described the dark green vapors that appear after the first fire has digested the two serpents for thirty days] is the Green Ladies of B. Valentine the beautifully green Venus & the green Venereal Emrauld & green earth of Snyders wth wch he fed his lunary ☿ & by vertue of wch Diana was to bring forth children & out of wch saith Ripley the blood of ye green Lyon is drawn in ye beginning of ye work. *Nam viridis et vegetabilis nostri argenti vivi substantia est Basilisci philosophici pabulum* saith Mundanus p 180. The spirit of this earth is ye fire in wch Pontanus digests his feculent matter, the blood of infants in wch y^{e} ☉ & ☽ bath themselves, the unclean green Lion w^{ch}, saith Ripley, is ye mean of joyning ye tinctures of ☉ & ☽ , the broth wch Medea poured on ye two serpents, the Venus by mediation of w^{ch} ☉ vulgar and the ☿ of 7 eagles saith Philalethes must be decocted & in wch ye same ☿ digested alone will give you the Philosophic Lune & ☉, & the Spirit of Grasseus where he saith y^{t} in the *via humida* the 4 Elem^{ts} become by steps one ☿ & this ☿ is divided into ☉ & ☿ wch must be reconjoyned by mediation of ye Spirit to give a third thing.[29]

Since he understood a chemical process to stand behind the imagery, he sometimes inserted bracketed explications in a text which translated it into prosaic chemistry. Such passages, if I may use a more recent image, have the effect of a dull railroad spike thrust into a swelling balloon of overheated gas. Our crude sperm flows from a trinity of immature substances, he wrote. Two [♂ & ♄] are extracted by the third [♁] and become a pure milky virgin substance from the menstruum of our sordid whore. You must draw the moon [sp^{t} of ♁] from the ferment [in distilling] and bring it from heaven to the earth [of ♂] and turn it into water and then into earth. Somewhat later in the same process, when the doves came flying and were enfolded in the arms of Venus and assuaged the green lion, he added the parenthetical explication, "(as alcalis do acids)."[30]

Newton did not accept everything he read in an alchemist at face value. After he recorded Sendivogius' assertion that he had revealed the extraction of philosophic mercury from Pontic water, Newton entered a blunt disclaimer: "he did not reveal the rectification of the water, the number of sublimations [, or] the proportion of water to Sulphur in any sublimation."[31] One set of notes he broke off and crossed out, and under them he set a judgment in tones that Hooke and Flamsteed would have recognized. *"Credo hic nihil adeptus."*[32] If he was convinced that the burgeoning imagery expressed a single work, he also learned early to distinguish the real adept from the pretender.

As he proceeded, moreover, and immersed himself more deeply in the art, he learned to use one authority to correct another. In *Ripley Expounded,* a composition of the 90's, Newton added a note to a passage listing three different proportions to use in calcination. "NB Philaletha in S.R. [*Secrets Reveal'd*] makes these three proportions belong to 3 works, but since Ripl. bids take wch you will of ye two first, those must belong to ye same."[33] A similar note followed a discussion from Generosus of a solution black as ink. "Whence (by comparing him with Ripley, Philaletha, & Scala,) it appears that Ripley omits all the first gross preparation & begins his gates wth the decoction of ☉ & ☽ in ☿ first to a black calx & then to a black water wch are ye two first gates."[34] Now and then, rather significantly I think, such passages implicitly cited Newton himself by failing to cite anyone else. "But these imbibitions seem not necessary to ye philosophic work," he wrote in reference to Mundanus' method of imbibing earth with animated spirit, "because the stone may be made without the two spirits & multiplied by them & the fixt salt (as was shewed above,) & lastly fermented by fusion."[35] No one could study the alchemists that long and that intensely without beginning to see himself as one of them.

When one follows the papers chronologically, he cannot fail to note a new tenor that begins to appear about 1680. In some limited sense, the majority of the papers continued to be reading notes; that is, they consisted mostly of citations from alchemical works. Reading notes they can hardly be called, however. Carrying the comparison and collation of texts to the next level of sophistication, Newton began to compose alchemical writings himself. As his very conception of the tradition demanded, he drew his compositions from the writings of earlier adepts and presented them as expositions of the single work to which the whole invisible fraternity, living and dead, testified. Nevertheless, I cannot imagine what these papers are if they are not his own treatises. It is simply incorrect to say that Newton never composed an alchemical treatise. In fact, he composed a considerable number.

An autograph sheet now found with *Keynes MS. 30,* although it apparently belongs with *Keynes MS. 35,* contains a list of 17 titles which sound like the titles of chapters—"In what manner metals are generated and corrupted in the veins of the earth," "On the seed, sperm, and body of minerals," "On the green lion," "On double mercury," and the like.[36] *Keynes MS. 35* does in fact contain a number of pieces which correspond to these titles, the first three of them frankly labelled "Cap. 1," "Cap. 3," [sic] "Cap. 3" with titles nearly identical to the first three on the list. Composed from the writings of earlier alchemists, the chapters attempted to bring together into a coherent exposition what they had said on the topics. Later items in *Keynes MS. 35*, while they lack the title of "Chapter" and in one case at least, merely collect excerpts (*"Ex Turba," "Ex Artephio,"* and so on) correspond nevertheless to headings on the list.

In much the same hand, which I place around 1680, Newton composed a paper labelled *The Regimen.* It began initially with six and was then expanded to seven "Aphorisms" which stated in brief the essentials of the work.

Aphorism 1.

The work consists of two parts the first of wch is called the gross work & by many imbibitions & putrefactions purges the matter from all its gross feces & exalts it highly in vertue & then whitens it.

Aph 2

The second part also putrefies the matter by severall imbibitions & thereby purges it from ye few remaining feces & exalts it much higher in vertue & then whitens it. For the two parts of the work resemble one another & have the same linear process. . . .

Aph 4

In both works the Sun & Moon are joyned & bathed & putrefied in their proper menstruum & in the second work by this conjunction they beget the young king whose birth is in a white colour & ends the second worke . . .

Aph 6

The young new born king is nourished in a bigger heat with milk drawn by destillation from the putrefied matter of the second work. With this milk he must be imbibed seven times to putrefy him sufficiently & then decocted to the white & red, & in passing to ye red he must be imbibed once or twice wth a little red oyle to fortify ye solary nature & make the red stone more fluxible. And this may be called the third work. The first work goes on no further then to putrefaction the second goes on to ye white & ye third to ye red.

Aph 7

> The white & red sulphurs are multiplied by their proper mercuries (white & red) of the second or third work, wherein a little of the fixt salt is dissolved.

"This Process," Newton continued, "I take to be ye work of the best Authors, Hermes, Turba, Morien, Artephius, Abraham ye Jew & Flammel, Scala, Ripley, Maier, the great Rosary, Charnock, Trevisan. Philaletha. Despagnet." Four long folios, filled with "Annotations," drew upon the writings listed to support the aphorisms.[37]

It cannot have been long after the composition of the *Regimen* that Newton began his most remarkable exercise in alchemical scholarship, the *Index chemicus.* First he drew up a list, Index 1a, of 115 headings in rough alphabetical order—a line for headings beginning with A, one for those beginning with B, and so on. He copied the headings, now precisely alphabetized, onto two folded sheets, Index 1, eight pages in all, with about half an inch between headings in which to enter references. As he added the references, he added other headings as well, and the list had grown to 251 entries when the exhausted space forced him to start anew with Index 2. He followed the same procedure as before, setting down the crowded headings of Index 1 at expanded intervals on twenty-four pages. By the time it had grown to 714 headings, Index 2 would hold no more, and Newton undertook to expand it again. A number of references to Mundanus, all apparently late additions, indicate that Index 2 was not outgrown before 1686. Meanwhile the *Index* was changing its character as well as its size. Initially it appears to have been a device to order the immense volume of information Newton was gathering. It was always a selective rather than an inclusive index, however; some references in Index 1 were dropped from Index 2, for example. More importantly, some of the entries were beginning to take on the form of small essays. As he prepared to expand the *Index* again, Newton composed drafts of some of the essays on separate sheets, and when he abandoned a version of the new index (Index 3a) before he had completed the A's, he treated what he had written as a draft for the final *Index chemicus* (or Index 3). A reference to the *Chymical Wedding,* added late, indicates that Index 3 was not completed before 1690, if indeed it was ever completed and closed.

In its ultimate form, the *Index chemicus* stretched beyond 100 pages and contained 879 headings. The 46 largest entries, filling about 42 pages, contain 1975 page references to 144 separate treatises and 100 different authors. By extrapolation, the entire *Index* must contain nearly 5000 separate references. Newton had combed all the major published collections—*Theatrum chemicum, Ars aurifera, Musaeum hermeticum, Theatrum chemicum Brittanicum,* and *Aurifontina chymica.* He had digested the various works of major writers such as Arnold of Villanova, Raymond Lull, Nicholas Flammel, Johann Grasshoff, Michael Maier, and Eirenaeus Philalethes (whose works were easily the most frequently cited). I spent a week dissecting the *Index* to the extent that I have. It is hard for me to imagine that anyone could have composed it in less than a thousand times that week—although I hasten to add that I cannot find room in Newton's career for any period approaching a thousand weeks.[38]

What purpose was the *Index chemicus* intended to serve? Even in its final form some of the entries contained only the names of books and page numbers. For the most part, however, the *Index* continued the task of expounding the one alchemical work behind the welter of images. "*Materia prima* is that which has been stripped of every form by putrefaction so that a new form can be introduced, that is, the black matter in the regimen of Saturn." It is also called hyle, chaos, the dark Abyss, and forty-five other names which Newton duly listed.[39] The essay under "Typho" explored the meaning of the term in four books of Michael Maier.[40]

The intensity of study required to produce the essay recalls Newton's eighty-eight closely written pages of notes on Maier and indicates how thoroughly he digested his books.[41] Many entries frankly speak, not to the compiler of the *Index,* but to an understood user of it. *Aes* is generally used to designate sulphur, he explained. Alchemists use *Altitudo* for mercurial water, and (analogously) they use *Duenech, Fumus albus, Ignis pontani, Leo viridis,* and many other terms to express aspects of the work which he explained. The drafts and revisions of essays assumed the same reader. Whether Newton expected him to be the owner of the *Index* in a printed book or a member of the invisible fraternity poring over a circulating manuscript even as Newton had done, I cannot tell. What does seem clear beyond reasonable doubt is the conclusion that the *Index chemicus* ended by being much more than Newton's private directory to passages on various topics.

Nor was the *Index* the last of his efforts to draw the alchemical tradition together into a coherent statement of the work. Three distinct compilations, at least one of which can be connected by the collation of references rather closely with the *Index,* either as source or as product, attempted to expound the work in much the style of the earlier *Regimen* but at much greater length. One of the compilations, which can be traced through four versions, addressed itself rather directly to alchemical practice. *The method of ye work,* the earliest version that I have identified, expounded Didier's procedure as it was published in the *Six Keys* (1688) and demonstrated that his procedure was identical to that of eleven other alchemical writers.[42] *Of ye first Gate* expounded the same procedure from Ripley and demonstrated that his practice was identical to that of Didier and a number of other alchemists, mostly but not entirely identical to the group cited in *The method of ye work.*[43] Newton utilized whole passages from the *First Gate* in the final version of this treatise, *Praxis,* which itself went through two drafts.[44] Another compilation, intended perhaps to be entitled *Decoctio,* survives in two drafts, and in related papers labelled *The Regimen* (which are distinct from the earlier *Regimen*).[45] In passing, let me remark on a correction in one of the *Regimen* papers which illustrates how coping with dragons and lions and doves of sundry hues only hints at the woes that beset the student of the art. The first menstruum ferments with the two dragons, Newton wrote, until it becomes "a dark red or yellow colour wth black & blue spots." Fearing perhaps that the description was too explicit, he crossed it out and wrote instead, "an amiable green dark colour."[46] If we frequently see Newton behind the alchemist in these papers, at least as frequently we see the alchemist in front of Newton.

The most extensive of the compilations, and the one most closely related to the *Index,* carries no general title.[47] In one version it was divided into six *Opera,* although further chapters exist which seem to be related. At least six drafts of one of the *opera, Extractio et rectificatio animae,* survive, four of *Extractio et rectificatio spiritus,* and at least two of three others. The description in the Sotheby catalogue of another manuscript which cannot be located now seems to point to other drafts as well.[48] Lest one mistakenly confuse the method of exposition through earlier writings for mere reading notes, Newton, who must have known his favorite authorities by heart, frequently left empty spaces after the cited titles where he intended later to insert page references.[49] In the latest versions, he made their status as treatises still more clear by moving the references into the margins where they assumed the role of footnotes, which, as every historian knows, are quite a different thing from reading notes. And the author of the recent *Principia* put his definitive mark on the composition when, after expounding the extraction of the soul from the philosophic chaos, he began a re-exposition of the process with the mathematician's *Idem aliter.*[50]

Successive drafts of these chapters reveal a significant development. Early in his career as alchemist, Newton had looked to Eirenaeus Philalethes for his most important source. Philalethes continued to be the most frequently cited author in the *Index.* During the early 90's, however, Newton turned increasingly to earlier authorities, not to contradict Philalethes, but to support him with more ancient testimony. Reading notes from Hermes' *Tabula smaragdina,* Ripley and Lull, and notes on Maria the Jewess survive from the 90's.[51] So also do several lists of the "best authors" with heavy emphasis on ancient, medieval, and Arab alchemists.[52] In similar fashion, successive drafts of the chapters in question here moved steadily toward an exposition of the work that relied primarily on medieval authorities.

Four other manuscripts from the 90's deserve brief mention. *The three fires* also expounded the single work to which all authors testify. There is only one page reference in the entire paper.[53] *Sendivogius explained* (if it was Newton's own composition) and *Ripley Expounded* followed the pattern of Philalethes' *Ripley Reviv'd,* on which Newton himself had drawn so heavily, describing the work by pretending to comment on an earlier alchemist.[54] There are also five chapters of a treatise without a general title which contain some revisions of the sort that abound in Newton's papers, and thus appear to be his composition.[55]

While he was reading and writing, Newton was also experimenting. Any assessment of his relation to alchemy must ultimately turn to the experiments, for if they reveal no signs of the art, the rest of the papers, whatever their arcane suggestion, will distill away as flatulent collections of outlandish imagery with no enduring substance. The earliest records, which appear to be, not notes of experiments as later ones were, but their summation, are in the hand of the late 60's and early 70's.[56] After trying several methods to extract the mercuries of metals, Newton focused his attention on a material which never thereafter departed far from the center of his chemical activity, the regulus of antimony. Metallic antimony, as we would say, the regulus was separated from stibnite (what the 17th century called antimony) either without the use of a metallic reducing agent to combine with the sulfur (in which case it was regulus per se) or with the use of a metal (in which case it was regulus of iron or copper, or whatever metal was used and was deemed apparently to embody that metal's active seed). Newton recorded precise details on how to obtain the whole gamut of reguli, and he added six rules for the experimenter to follow and several signs for him to observe. In particular he noted the proportions for regulus of Mars (as iron was called of course), 4 to 9, with a good smart fire; "ye Reg after a purgation or two starred very well," he added. The star that could appear on the surface of the antimony when the metallic veins so arranged themselves made it the star regulus, and the star regulus was fraught with alchemical significance.

Sometime around the mid 70's, Newton composed or transcribed a paper entitled *Clavis.* The paper carries no explicit mention of its author, and it is impossible to state with finality that it was or was not Newton's. Nevertheless, its intimate connection with Newton's notes on the preparation of the star regulus, and its use of the proportion 4 to 9 at which Newton arrived, make it extremely likely at the very least that Newton did compose it.[57] The sign of the star indicates, the author asserted, "that the soul of the iron has been made totally volatile by the virtue of the antimony." Since the regulus melts with gold and evaporates entirely away, it may be the sulfur of gold. Amalgamate it with common mercury which receives a spiritual semen from the regulus. "The spiritual semen is a fire which will purge all the superfluities of the mercury, the fermental virtue intervening." The mediation of the virgin Diana (pure silver), which the paper later referred to as the doves of Diana, was necessary in order for the two to unite. Digest the amalgam in a sealed vessel on a slow fire for several hours and then grind it in a mortar; the mercury spits out its blackness or feces, which can be washed off. After repeated purgings and washings the amalgam becomes like shining silver. Now distill it, amalgamate it anew, purge and wash, and distill again, repeating the whole operation seven or nine times.

> On the seventh time you will have a mercury dissolving all metals, particularly gold. I know whereof I write, for I have in the fire manifold glasses with gold and this mercury. In these glasses they grow in the form of a tree, and by a continued circulation the trees are dissolved again with the work into new mercury. I have such a vessel in the fire with gold thus dissolved, where the gold was visibly not dissolved by a corrosive into atoms, but extrinsically and intrinsically into a mercury as living and mobile as any mercury found in the world. For it makes gold begin to swell, to be swollen, to putrefy and to spring forth into sprouts and branches, changing colors daily, the appearances of which fascinate me every day. I reckon this a great secret in Alchemy . . .

Further support for Newton's authorship lies in the concept of mediation by Diana and the singling out of mercury sublimed seven times. This was the mercury of seven eagles that appeared repeatedly, with Diana and her doves, throughout Newton's alchemical papers.

I have suggested that the general treatises that Newton composed on the alchemical work used the writings of earlier alchemists to expound the one true process. His own experience at the furnace helped to determine his recognition of the process. Thus his references to mercury of seven eagles rested on his own experiments. One of the basic procedures which the treatises also repeated was the amalgamation of the regulus of Mars by the mediation of the doves in the work with vulgar mercury.[58] It was also typical of Newton and his relation of experience to tradition that a short paragraph in the *Clavis* referred the reader to Bernard of Trevisan's letter to Thomas of Bologna for an exposition of the operation.

Newton's experimental notes, many of which were dated, began about 1678.[59] With a few exceptions, they did not employ the imagery of alchemy, but on close inspection they reveal unmistakable marks which only an alchemist could have printed on them. One of the marks is found in the exceptions where Newton did use the terminology of the art—the serpent, the Caduceus, Saturn impregnated.[60] The tendency to use ores instead of metals reflected the consideration, frequently repeated in the papers, that the seed of metals must be sought among the living; metals that have undergone the fire are dead.[61] Saturn and lead appear constantly in the papers as the material of metal in its lowest stage of specification and hence a likely source of the metallic ens; Newton's experiments dealt with lead more than with any other metal except antimony, which was itself taken to be the child of lead.[62] Frequently Newton tested his materials, especially the *capita mortua,* by seeing if they would flow like wax on a sheet of iron or glass in the fire. The same test, ultimately related to the process of inceration appeared repeatedly among the alchemical papers.[63]

While these characteristics may be suggestive, all of them taken together cannot by themselves justify the conclusion that Newton's experiments were alchemical. They do help, however, in the interpretation of other features in the experimental notes. One of these is the "net," a material that found its way frequently into Newton's crucibles. He had learned to make the net before he began the dated notes.

> R ♂ 9¼, ♀ 4 gave a substance with a pit hemisphericall & wrought like a net wth hollow work as twere cut in.[64]

He tried two other proportions of regulus of Mars to Venus (or copper) before he decided that 4 to 8½ or 9 was the best. As the notes plainly indicate, the sheer physical suggestion of a network helped to identify the regulus as the net; somewhat later the pit and grain would help to identify a closely related regulus as the hollow oak. Nevertheless, the physical network was no more than the outward sign of an inward virtue, the copulation of Mars and Venus, the male and female principles central to every statemen of the art, the sun and moon from whose union, he said in Aphorism 6 of the *Regimen,* the new king is born. The symbol of the net derived from the myth that Vulcan, the ingenious (though not for that any the less cuckolded) spouse of Venus, had devised a net about her bed by which he trapped her in dalliance with Mars. In the *Praxis,* Newton referred to "♂ & ♀ wrapt in the net of Vulcan . . ."[65] In somewhat different though not inconsistent terms, he later suggested that the net was an alternative to Diana's doves as a mediator to unite the seed of Mars to Mercury.[66] Clearly the net was a powerful alchemical device, and it is not surprising that it showed up repeatedly in Newton's later summations of the work.[67]

As far as I have been able to discover, the "oak" first appeared in Newton's experimentation in 1682. He defined it there as "Reg ♂ ♀ ♁." I do not know how to distinguish this regulus from the net, and I have already suggested that the pit in the regulus he called the net may have suggested the hollow oak, as he sometimes called it. So also the grain of the metal may have suggested the grain of the wood, for after he had experimented a bit, imbibing the oak with salt of antimony and melting it, he obtained a white brittle product, on the grains of which he remarked. "And this I conceive to be ye right preparation of ye Oak," he concluded.[68] Again the external appearance was merely incidental. When he first used it, Newton imbibed the oak with vinegar of antimony and sublimed it both with vitriol and with sal ammoniac. The procedure powerfully evokes the imagery of the oak in Newton's alchemical treatises, where a serpent devoured Cadmus and his companions, but at length Cadmus fixed the serpent to a hollow oak.[69] On one occasion, Newton mixed the oak with sal ammoniac and vitriol and remarked that "ye matter sublimed liked Dragons teeth."[70] In other manuscripts, Newton described the hollow oak as a chaos of the four elements (♂ , ♃ , ♀ , ♁) and the quintessence (♆ , which usually represented Bismuth).[71] The rest of the symbolism, that is, Cadmus and his companions and the serpent, remained unchanged. An experiment on 26 April 1686 or shortly thereafter made a regulus from the chaos (compounded, as it were, by the addition of Saturn). "This Reg had a glorious starr," he noted. He proceeded then to melt it with copper (or Venus) and saltpeter and obtained the familiar pitted regulus "wrought wth network." "This metal broaken," he stated in conclusion, "looked finer then any other I ever made wthout ye ores. The next time proceed thus."[72] Clearly the oak belonged in the same general alchemical framework as the net. In his alchemical treatises the oak, fermented (once with two dragons) and distilled, yielded the blood of the green lion, our Venus, dry water, double mercury, or Artephius' third fire.

The early 80's were a period of intense alchemical experimentation for Newton, reaching a climax in the spring of 1681. A series of experiments investigating the volatizing virtue had culminated in one in which a sublimate of copper (*Venus volans*) had carried up lead.

> May 10 1681 I understood that the morning star is Venus and that she is the daughter of Saturn and one of the doves. May 14 I understood ⋲ [the trident?]. May 15 I understood "There are indeed certain sublimations of mercury" &c as also another dove: that is a sublimate which is wholly feculent rises from its bodies white, leaves a black feces, which is washed by solution, in the bottom, and mercury is sublimed again from the cleansed bodies until no more feces remains in the bottom. Is this very pure sublimate not ⊸*? [sophic sal ammoniac?][73]

Since the next experiment employed "sophic *," I am prepared to believe that Newton's special symbol, a deliberately distorted *, meant to express this concept.

> May 18 I perfected the ideal solution [*ideam solutionis*] That is, two equal salts carry up Saturn. Then he carries up the stone and joined with malleable Jupiter [as much as one wants to say "tin" here, Newton wrote out *Jove* instead of the symbol ♃] also makes ⚹ [sophic sal ammoniac?] and that in such proportion that Jupiter grasps the scepter. Then the eagle carries Jupiter up. Hence Saturn can be combined without salts in the desired proportion so that fire does not predominate. At last mercury sublimate and sophic sal ammoniac [* *praeparat*] shatter the helmet and the menstruum carries everything up.[74]

Whereas Newton set down his laboratory notes in English, he wrote these two paragraphs in Latin and put large X's through them, less to cross them out, it appears, than to call attention to them. There is a sheet in *MS. 3973* with the date July 10 but no year; the hand is compatible with 1681, and I take it to belong to the same climactic series. "July 10, I saw sophic sal ammoniac [* *Philosophicum*]. It is not precipitated by salt of tartar."[75] Beyond this sheet, nothing in *MS. 3973* corresponds to the paragraphs in notebook *3975* although many of its other notes were transcribed from *MS. 3973*. The paragraphs thereby take on significance as interpretations set down among the experiments.

Unfortunately, the interpretations need to be interpreted, and only Newton would be fully qualified to do it. Nevertheless, the imagery of the passages is familiar to a student of Newton's own treatises. Both in the *Praxis* series and in the series represented by the six *Opera,* Newton made much of the myth of Saturn and Jupiter, in which Saturn, thinking to eat his son Jupiter, eats a stone instead and vomits it up again; whereupon Jupiter deposes Saturn, commands peace with his scepter, and is carried up to his throne by an eagle.[76] I have suggested before that in his own treatises Newton used the alchemical tradition to expound the work as he himself had confirmed it at the furnace. If I am correct, the experiments of the early 1680's laid the foundation for the treatises written five to ten years later. Certainly the imagery of the paragraphs above echoed and re-echoed in the treatises, and with it went the distinction, implicit in the paragraphs, between sal ammoniac of the vulgar and sal ammoniac of the philosophers or our sal ammoniac.[77] "I am ye earthly black eagle and was heretofore washed wth ye corrosive of Neptune and by ye Venereal property exalted into a most beautiful crystalline weighty essence . . .," he copied from de Monte-Snyders at about the time of the experiments. "I am the Eagle, I am ye P[hilosop]hick Air, & am also ye true Salarmoniack. I by my wings have brought ye king of ye Earth up to his throne . . ."[78] In fact, one facet of the vision remained to be confirmed. "Friday May 23 [1684] I made Jupiter fly on the wings of the eagle."[79]

I am neither a chemist sufficiently learned nor an adept sufficiently exalted to penetrate very deeply into these notes, and I would not be understood as offering an explication of them. I am striving for a less ambitious goal—to show (which is less than to demonstrate) that Newton's own chemical experimentation pulsed with alchemical overtones, that it was indeed alchemical experimentation. A basic concern with volatizing metals animates it all. At least among such 17th century alchemists as Sendivogius and Philalethes, both of whom influenced Newton greatly, two logically independent themes wove themselves together to form the fabric of the work—on the one hand, generation by male and female as a process universal in nature, in the mineral kingdom as well as in the animal and vegetable; on the other hand, the need to purge and purify in order that the spiritual seeds of things can attain their ends. The conflation of these themes could produce some striking imagery on occasion. "I say our true Sperm flows from a Trinity of Substances in one Essence,"

Philalethes wrote, "of which two are extracted out of the Earth of their Nativity by the third, and then become a pure milky Virgin-like Nature, drawn from the Menstruum of our Sordid Whore."[80] The sulfur or male principle was the spiritual seed which needed to be cleansed of inhibiting feces before it could properly vivify its body; initial separation of the two principles and the purification of both were necessary steps toward their eventual perfect unification. When Newton referred, as he did at least once in his notes, to spiritualizing metals, he implicitly imposed an alchemical interpretation on the entire enterprise.[81] The alchemical interpretation in turn casts new light on his meaning when he discussed the generation of air from metallic fumes in his letter to Boyle and in the essay called *De aere,* and those two pieces in turn illuminate the meaning of the experiments.[82]

When the experimental notes are taken with the reading notes and treatises, the corpus of papers that Newton left behind begins to appear as a treasure unique in the entire history of alchemy. On the one hand, it is reasonable to surmise that alchemy has never had a more informed and perceptive student. For long years he pored over the records of the tradition, collating and comparing, and he left behind in the *Index* and elsewhere detailed expositions and interpretations of its meaning. At the same time, he pursued alchemy at the furnace, and in his experimental notes, largely expressed without alchemical imagery, he translated the symbols into mundane chemicals. In conjunction, the two sets of papers would seem to offer an unparalleled instrument to penetrate the mysteries of the art. I can only sigh in disbelief that historians of alchemy have left this unique legacy largely unclaimed.[83] Perhaps they cannot wholly believe that such a bequest has descended from the fountainhead of the modern tradition of positive science. Who, indeed, would have expected the author of the *Principia,* within six years of its publication, to claim that he had the secret of multiplication?

> Thus you may multiply each stone 4 times & no more [he wrote in the climax of the *Praxis*] for they will then become oyles shining in ye dark & fit for magicall uses. You may ferment them w^th ☉ & ☽ by keeping the stone & metall in fusion together for a day, & then project upon metalls. This is the multiplication in quality. You may multiply it in quantity by the mercuries of wch you made it at first, amalgaming ye stone with y^e ☿ of 3 or more eagles & adding their weight of ye water, & if you designe it for metalls you may melt every time three parts of ☉ wth one of ye stone. Every multiplication will encrease it's vertue ten times &, if you use y^e ☿ of ye 2^d or 3^d rotation wth out ye spirit perhaps a thousand times. Thus you may multiply to infinity.[84]

Newton's last dated experiments fell in February 1696.[85] Less than two months later, he was Warden of the Mint. One cannot avoid wondering if the Lords Commissioners of the Treasury were fully informed about their new public servant.

Had they been fully informed, surely the Lords Commissioners as men of their age would have concluded that the newly appointed Warden was an alchemist. And so indeed I think anyone who is fully informed about his papers must conclude. In contrast to the Commissioners, however, we are not men of the 17th century, and merely to close the issue by judging Newton an alchemist is not very helpful. Alchemy was no one thing. By Newton's day the art had been pursued consistently in the West for nearly two thousand years in a series of widely divergent cultures. The alchemy of Maria the Jewess had not been that of Raymund Lull, nor had Lull's been that of Sendivogius. Inevitably the alchemists of every age had expounded the work in terms comprehensible to them. In the Medieval alchemists, for example, Newton found a process embodying Aristotle's four elements and the possibility of their mutual conversion, and his first impulse, which he clearly shared with other 17th century adepts, was to translate their expressions into a more familiar philosophy of nature.[86] Like the Medieval alchemists, Newton also practiced the art in a given intellectual context, in which alone his activity can become comprehensible. We must go beyond the conclusion that Newton was an alchemist and try to place his alchemy in the framework of his scientific career.

Immediately, the chronological pattern of his hermetic activity is suggestive. As I indicated before, his alchemical activity neatly bracketed the *Principia,* rising steadily in intensity in the late 70's and early 80's and, with scarcely an interruption for the *Principia,* reaching a quantitative climax (or perhaps merely continuing unabated) in the early 90's. The *Principia,* in turn, among its many facets, proposed a radical revision in the prevailing mechanical philosophy of nature by embracing the notion of action at a distance as the central aspect of its conceptual scheme. The question arises: Can these two developments have been related?

Let us turn back initially to the beginning of Newton's scientific career when, as an undergraduate, he discovered the new world of the mechanical philosophy. In notebook *3996,* in the *Questiones quaedam Philosophiae,* Newton recorded his intellectual awakening as he digested the writings of Descartes, Gassendi, Charleton, and Boyle.[87] But with the orthodox mechanical philosophers, he also discovered a dissenting school represented for him initially by the Cambridge Platonist, Henry More. More welcomed the mechanical philosophy for its clear distinction of body and spirit. For him, the distinction served a religious end, for by showing the limitations of matter, it demonstrated the necessity of spirit. His reservations about the Cartesian philosophy centered on its tendency to exclude spirit from the ordinary operations of nature. On the cosmic plane, it proposed a universe that ran by mechanical necessity alone, whereas More was determined to save the immanent activity of God. On the lower plane of particular phenomena, More was convinced that certain events, above all the phenomena of organic life, display at once intelligent planning and spontaneous activity that could never result from matter in motion alone. Matter is stupid and passive; organized living beings are active and they display the results of intelligent planning, two characteristics that can be traced, in the cases of men and animals, to a single source called a soul. In general terms, the passivity of matter, the object of the mechanical philosophy, requires the presence of active principles to complete, perfect, and animate it. Notes from Henry More appear with those from Descartes in the *Quaestiones.* They suggest that from the beginning of his career Newton had reservations about central aspects of the mechanical program.

Not long after his undergraduate days, Newton composed an essay, *De gravitatione et aequipondio fluidorum,* which has only recently been published. *De gravitatione* is a complicated treatise. At least two arguments take place in it, as indeed they did already in the *Quaestiones*—on the one hand, Gassendist atomism against Cartesian philosophy; on the other hand, the active principles of neoplatonic philosophy against the passivity of mechanical matter. Only the latter concerns us here. Newton insisted that body must be distinguished from space. Space, or extension, is immobile and incapable of inducing any change in the motion of a body or the thought of a mind. Of bodies exactly the contrary is true; endowed with two basic faculties, they can move other bodies and stimulate perceptions in minds. These faculties derive from the activity of God. For the existence of bodies, Newton decided, there are required "extension and the activity [*actus*] of the divine will . . ." *De gravitatione* also insisted on the analogy between the activity of God, uncreated spirit, and the activity of soul, created spirit, an analogy which remained a permanent characteristic of Newton's natural philosophy. God created the world by the action [*actione*] of His will; we move our bodies by the action [*actione*] of our wills.[88]

The general discussion of natural philosophy in *De gravitatione* introduced specific questions in mechanics. A few years earlier, in the *Waste Book,* Newton had embraced the Cartesian conception of motion, in effect, the principle of inertia although it did not carry that name. The principle of inertia, however, was the central member of the Cartesian world machine which *De gravitatione* rejected. Newton solved his problem by converting the Medieval concept of impetus into an active principle. In his discussion of motion, he insisted on the equation of impressed force with every change of motion. At the end of the introduction he summarized his doctrine in a number of definitions.

> Definition 5. Force is the causal principle of motion and rest. And it is either an external one that generates or destroys or otherwise changes impressed motion in some body; or it is an internal principle by which existing motion or rest is conserved in a body, and by which any being endeavours to continue in its state and opposes resistance. . . .
>
> Definition 7. Impetus is force in so far as it is impressed on a thing.
>
> Definition 8. Inertia is force within a body, lest its state should be easily changed by an external exciting force.[89]

The universe of *De gravitatione* was composed of passive matter informed by active principles.

Here was a view of nature that any alchemist would recognize. A man already disposed to see nature in such terms was likely to respond to the rhythms of the art with its active spirit or sulphur or seed hidden in a mass of encumbering feces from which it must be cleansed. Newton's papers abound with similar sentiments copied out of the masters he read.[90] At much the same time that he composed *De gravitatione,* he also turned his hand to an alchemical essay, probably the first such that he wrote, in which the same theme of active and passive prevails. More a sketch than a finished treatise, it has no name but is generally known as *The Vegetation of Metals.*[91] A number of passages in it foreshadowed the *Hypothesis of Light* of which it constituted a distantly related early draft. Like the *Hypothesis,* it combined mechanistic elements, particles of matter in motion, with features that transcended mechanism. *The Vegetation of Metals* started with a set of twelve statements which, with the brief sketches of argument that accompany some of them, look like chapter titles for an intended treatise.

> 1. Of Natures obvious laws and processes in vegetation.
> 2. That metalls vegetate after the same laws.

Statement 6, further down the page, summed up the central theme of the essay in stating that vegetation is the effect of a spirit which is "ye same in all things," differing only in its degree of maturity and in the rude matter it animates. Newton described how this spirit, rising from the center of the earth, concreted in water to form the various substances in the crust of the earth, and was by this so alienated from its metallic nature that these substances hinder the alchemical work. Significantly, he insisted that the alienation stemmed from the fact that the change the spirit underwent as it concreted in water to form salts and minerals was not a process of vegetation but "a gros mechanicall transposition of parts." If the spirit can be freed from salts by chemical means, it can recover "metallick life & . . . pristine metalline forme." The talk of metallic fumes as active spirits in nature recalls Newton's own later experiments in volatizing metals.

Some of the metallic exhalations, *Vegetation* continued, escape into the air and beyond that into the aetherial regions, and by their rise compel an equal quantity of aether to descend. Newton likened the circulation to the breathing of an animal, which the earth resembles, drawing in refreshment and vital ferment and breathing out gross exhalations.

> This is the subtil spirit [he said of the aether] . . . this is Natures universall agent, her secret fire, ye onely ferment & principle of all vegetation. The material soule of all matter wch being constantly inspired from above pervades & concretes wth it into one form & then if incited by a gentle heat actuates & enlivens it . . .

> Note that tis more probable ye aether is but a vehicle to some more active spt. . . . This spt perhaps is ye body of light becaus both have a prodigious active principle both are perpetuall workers.

Although he did not explicitly equate it with alchemical sulfur, the ultimate kernel of "our gold," he made the identity sufficiently clear.

At the end of *Vegetation,* in a discussion of putrefaction, the active-passive dichotomy offered a general distinction between nature's two modes of action, mechanical and vegetable. The principles of nature's vegetable actions are seeds, "her onely agents, her fire, her soule, her life." The seed is that portion of any substance that has attained full maturity; vegetation is "ye acting of w^{t} is most maturated or specificate upon that wch is less specificate or mature to make it as mature as it selfe." The seed is never more than a tiny part of the whole surrounded by "dead earth & insipid water." These grosser substances function as vehicles in which the seeds perform their actions, and the gross substances take on different appearances as their particles are moved about. Such changes, Newton asserted, are purely mechanical, like the mixing of two colored powders to get a third color, or the coagulation of milk into butter by mere agitation. All of the operations of vulgar chemistry, however impressive their display, are nothing but mechanical conjunctions and separations of particles. We must assume that nature employs the same means for the same effects.

> But so far as by vegetation such changes are wrought as cannot bee done wthout it wee must have recourse to som further cause And this difference is vast & fundamental because nothing could ever yet bee made wthout vegetation wch nature useth to produce by it. . . . There is therefore besides ye sensible changes wrought in ye textures of ye grosser matter a more subtile secret & noble way of working in all vegetation which makes its products distinct from all others & ye immediate seat of thes operations is not ye whole bulk of matter, but rather an exceeding subtile & inimaginably small portion of matter diffused through the masse wch if it were seperated there would remain but a dead & inactive earth.[92]

Among its leading artists, alchemy was always much more than goldmaking. It was a philosophy that claimed to offer the ultimate insight into the workings of nature. With the *Vegetation* in mind, perhaps we can understand the thrust of John Collins' uncomprehending report in 1675 that Newton was "intent upon Chimicall Studies and practises, and . . . beginning to thinke mathcal Speculations to grow at least nice and dry, if not somewhat barren . . ."[93]

The active-passive dichotomy expressed itself in *The Hypothesis of Light* in 1675, both in presenting the aether as the active spirit in nature, and in suggesting a repetition of the dualism on a higher plane within the aether itself. The aether, Newton said, is not one uniform substance; it is "compounded partly of the maine flegmatic body of aether partly of other various aethereall Spirits . . ."[94] Nearly ten years later, when he began to compose the *Principia,* he started initially, not from the principle of inertia, but from the mechanics of *De gravitatione,* in which the motions of bodies are determined by the interaction of external forces with the forces internal to bodies, active principles animating lifeless matter which plays no positive role whatever.

During that same decade before the *Principia,* Newton devoted himself overwhelmingly to alchemy. His experimental notes inform us that over the furnace he continually confronted active principles, chemical analogues of the spontaneous activity in nature that had convinced philosophers such as Henry More of the necessity of active principles to animate inert mechanisms. The very theme of the *Vegetation of Metals,* repeating the alchemical conviction that metals procreate by male and female, suggests how close the analogy appeared. The experimental notes abound with active verbs of a sort that are hardly met in a chemical text of the 20th century. When spelter was added to a solution of aqua fortis and sal ammoniac, "ye menstruum wrought upon ye spelter continually till it had dissolved it." The solution "boyled vehemently so as to send out a long white blast like an Aeolipile wth great force & by ye vehemence of ye motion to crack ye neck of ye glass." A similar solution of tin "fell a working wth a sudden violent fermentation . . ."[95] The heat generated in such operations impressed Newton; his notes remind us why the alchemical active principle was called a sulfur or fire. Sometimes similar passages employ other verbs. The spirit "draws" or "extracts" the salts of metals, recalling the frequent use of attractions and alchemical magnets among the authors he read.[96] When one substance combined with another in Newton's crucibles, it "laid hold" on the other; if the two sublimed together, one "carried up" the other; and if they did not sublime, one "held down" the other.[97] He observed that the amount of one substance on which another could lay hold was limited, and with that amount it was satiated. "If the metall be not too much," he concluded from one experiment, "it [a glassy scoria above a regulus] eats it all if more then it can eat it lets ye rest fall down."[98] He also noticed that one substance did not work on every other substance, but in general only on those to which it was related—the principle that like joins with like, which was common in alchemical literature.[99] Substances of contrary natures would not work on each other, although, as he had noticed already in the *Clavis,* a third substance sometimes mediated between contraries, enabling them to combine.[100] And finally the fact that certain salts deliquesced, that they laid hold on vapors in the air and "ran *per deliquium,*" seized Newton's attention.[101] In his eyes, the capacity to deliquesce revealed the presence of sophic sal ammoniac in a substance.

All of these types of chemical activity are familiar to students of Newton. In Query 31 they furnished the principal substance of the argument that particles of matter attract and repel each other. Indeed, he called the forces of attraction and repulsion "active principles," contrasted them with the passivity of inert matter, and in an argument reminiscent of Henry More and Ralph Cudworth, asserted that without such active principles "the Bodies of the Earth, Planets, Comets, Sun, and all things in them, would grow cold and freeze, and become inactive Masses; and all Putrefaction, Generation, Vegetation and Life would cease, and the Planets and Comets would not remain in their Orbs."[102]

Newton published Query 31 in 1706, but twenty years earlier, as he completed the *Principia,* he composed a shorter version of the same argument for a *Conclusio,* which he finally suppressed. The concept of forces between particles of matter, the basic concept both of the *Principia* in particular and of Newtonian science in general, grew from more than one root. Both in the *Conclusio* and in Query 31, Newton supported it by reference to the cohesion of bodies, capillary phenomena, surface tension in fluids, and the expansion of gases. These examples had furnished central features of his earlier speculations, and they trace their history in Newton's thought back to the *Quaestiones* in his undergraduate notebook. Some years before the *Principia,* according to his own account, he had convinced himself by an experiment with a pendulum that the aether does not exist.[103] Beyond its alchemical role, the aether had functioned as a mechanical medium that caused a number of phenomena, such as gravitational, electrical, and magnetic actions, which he now referred to attractions. Although such considerations contributed, the fact nevertheless remains that chemical phenomena carried the burden of the argument in the *Conclusio* as they did in Query 31, and with no exception that I have found the chemical phenomena cited in the *Conclusio* appeared in his own alchemical experiments. Newton's conception of matter as he now expounded it provides a subtle reminder of the role of alchemy. Bodies, he argued, are strings of particles joined together in tenuous nets. *Particulas retiformes*— the phrase, which appeared repeatedly in his discussions of matter around 1687, evokes the alchemical net that clasped Venus in the embrace of Mars, the basic union of the active and passive principles which in an altered form remained the foundation of Newton's conception of matter.[104]

"I may say briefly," he stated in the first paragraph of the *Conclusio* as he introduced the argument for short range forces between particles, ". . . that nature is exceedingly simple and conformable to herself. Whatever reasoning holds for greater motions, should hold for lesser ones as well. The former depend upon the greater attractive forces of larger bodies, and I suspect that the latter depend upon the lesser forces, as yet unobserved, of insensible particles."[105] The structure of the argument supports the existence of short range forces on the analogy of gravity. When we recall where Newton's attention was fixed during the decade before 1685, we are apt to conclude that the relation of the two was exactly the opposite. The forces of attraction and repulsion between particles of matter, including gravitational attraction which was probably the last one to appear, were primarily the offspring of alchemical active principles.

They were offspring transformed, however. The alchemical active principle was an entity capable of being separated from the mass it animated. It was a seed to be planted in passive soil, a yeast to be added to dough. Newton's forces could have no separate existence. Their indivisible union with matter reflected the final goal of alchemical manipulation, the perfect union of the two principles in the stone. In Newton's mature natural philosophy, however, the union was not effected by the scientist from independent entities; removed beyond the possibility of dissolution, it was a brute fact of nature effected by God in the creation. Newton never rid himself completely of the earlier concept of active principle. Sulphur remained a special substance in his eyes. He found, for example, that refractive power is proportional to density in transparent media, except for those with a high content of sulphur, which have proportionally larger refractive indices.[106] *De natura acidorum* related the activity of sulfur to the acid it contains. "For whatever doth strongly attract, and is strongly attracted, may be call'd an Acid."[107] On at least one late occasion he repeated the notion found in the *Vegetation* that bodies receive their "most active powers" from the particles of light, "the most active of all bodies known to us," that enter their composition.[108] Such passages appear to me as deposits from an earlier period, materials not fully assimilated in the further stage to which his thought had progressed. Newton's experimental notes from which I have quoted suggest how the details of chemical activity could have helped effect the conversion of the concept of active principle to its new form. The very composition of the *Principia* may also have worked to bring the new concept into the full light of consciousness. The essence of an attraction in contrast to an active principle, was its quantitative definition whereby it might fit into the structure of rational mechanics. Newton did not command a sophisticated mechanics when he wrote *De motu* late in 1684. The elaboration of his mechanics during the next few months made use of the idea of attraction and could well have influenced that idea in return. A sentence in *De Natura acidorum,* which catches both concepts of active principle in a fleeting embrace, lets us glimpse the mutation in the moment of change as it were. "The Particles of Acids . . . are endued with a great Attractive Force; in which Force their Activity consists; and thereby also they affect and stimulate the Organ of Taste, and dissolve such Bodies as they can come at."[109] To such estate had sunk the serpent that devoured Cadmus and his companions and was fixed to an hollow oak.

In the relation between Newton and alchemy, the influence did not pass all one way. If alchemy influenced Newton, so Newton influenced alchemy. How indeed could it have been otherwise? Could any activity undergo thirty years of intensive study by genius of that order and remain unaffected? One of the characteristics that has caught the eye of everyone who has looked at his experimental notes is their quantitative precision. The same spirit affected his study of alchemical texts. By far the longest entry in the *Index chemicus* was under the heading *Pondus.* Newton filled nearly four pages with all the quantitative relations he could find in extant alchemical literature.[110] Similar quantitative concerns fill the record of his own experiments. In August 1682, he investigated the proper proportion of sublimate of antimony to sublime from lead ore impregnated with salt of antimony. First he tried three samples of impregnated lead ore of 6 grains each, adding to them 10, 12, and 14 grains of sublimate of antimony. When he was not able to distinguish between the residues (all of which weighed 6 grains), he tried it again with larger amounts, at first three samples and then six of 60 grains each. To them he added 60, 80, 100, 120, 150, and 180 grains of sublimate. The first left a residue of 56 grains; each successive one was slightly smaller down to the last which weighed 52 grains. He tried how the residues flowed on a red hot iron; the last three all flowed and spread very well. On a hot sheet of glass, the last was wholly transparent, the fifth nearly so with a little whiteness in it, and the fourth more opaque. Taking all the evidence together, he concluded that something between the fourth and the fifth proportions, either 2 to 1 (the fourth) or at most 9 to 4 was to be used.[111] Not only did he weigh his samples and his products; he also started measuring specific gravities. Three samples of antimony compared to water as 4.27, 3.96, and 4.27.[112]

In all the experiments he used extreme precision. Paintings of alchemists have created an image like Macbeth's witches stirring huge caldrons of molten brew. Newton's notes present quite a contrasting image—of a modern technician clothed in a white coat and handling pipettes. Very rarely did he employ more than a few ounces of ingredients, and frequently he dealt with tiny fractions of an ounce, such that a candle sufficed for his heat. An experiment in 1683/4 sublimed four different proportions of a salt made from sophic sal ammoniac and vitriol from a precipitate of the net imbibed with antimonial vinegar. The total amounts of all cases weighed less than 20 grains (in a system of 480 grains to an ounce). The first three attempts left residues of 9½, 3 1/7, and 3 1/6 grains, the fraction in each case measuring the amount by which the salt fell short of carrying up its own weight. On the fourth trial, 5 grains of precipitate and 4 of salt "left 1^{gr} exactly of a sweetish & a litle stiptick tast but not strong. This was tryed wth great exactness & caution, the matters being mingled on a looking-glass that none of them might be lost. The last but one was also tryed carefully enough ye matters being mingled on a glass wth ye point of a knife."[113] Some may wish to doubt that he could distinguish fractions of a grain; and since the results cannot be compared readily with modern data, his degree of precision cannot be checked. Others may recall his precision in quite another field, his measurements of colored rings. He insisted on fractions of a hundredth of an inch in those measurements, and his results computed from the measurements compare favorably with the wave-lengths modern optics employs. For myself, I am unable to conceive of that deadly serious countenance deliberately deluding itself.

Hand in hand with the quantitative procedure, a method of analysis developed that was independent of the goals of alchemy and looked forward toward Lavoisier. The knowledge that sal ammoniac ran *per deliquium* became a test for the presence of sal ammoniac, or perhaps the volatising virtue, which he regularly recorded. "Sublimate of ♀ & ☿ kept two months in winter in a white paper," he noted on one occasion, "made ye paper very wet & also wetted other contiguous papers & pouders kept in them."[114] A stiptic taste also indicated the presence of sal ammoniac or volatising virtue. He had tests of other sorts. On at least one occasion, after using water to precipitate an antimonial sublimate, he tested the water with salt of tartar; when it precipitated nothing more, he concluded that the water had precipitated all of the antimony.[115] One sheet among his papers contains a list of about thirty precipitations. Oil of salt precipitates silver from aqua fortis. Iron, salt of tartar, and saltpeter precipitate antimony. Sal ammoniac precipitates silver and also lead. Copper precipitates silver, lead copper, tin lead. With most of the displacement reactions that ultimately appeared in Query 31, the list was systemized as though to serve as an analytic tool.[116] Occasionally something resembling quantitative analyses appeared.

> NB. White sublimate of mineral ♁ 17^{gr} holds * 12^{gr}, & ♁ 5^{gr}: & therefore must [be] joyned wth Iron ore 4^{gr} to make the ♁ to y^{e} ♂ as 2 to 1 For 4^{gr} of ♂ holds 4/3 of fixt scoria & 8/3 of volatile matter whereof 7½/3 is pure metall & ½/3 impure 🜍. So then ♂ must be to sublimate of ♁ as 4 to 17 or 1 to 4¼. But note that the matter wch remained below after sublimina-

> tion was totaly dissolvable in water, & therefore voyd of ♁ & by consequence all y^{e} ♁ ascends in ye 1st sublimation. Yet ye remaining matter melts wth * into a black water & in good quantity ascends. Coroll. Hence tis best to make sublimate of ♁ to ♂ as 3 to 1.[117]

On another occasion he dissolved antimony in a solution of aqua fortis and sal ammoniac, carefully measuring all the quantities of course. Then he precipitated the antimony (or *mercurius vitae*) with fresh water and filtered and distilled the solution. After the flegm, a dry white salt came over and left a saltish feces behind. From the measurements he concluded that the water, aqua fortis, sal ammoniac, antimony, volatile salt, and feces were to each other as 64, 32, 16, 25, 5, and 1.[118]

Surely alchemy had never known anything like this before. It was indeed more than alchemy could survive. The active principles that Newton met in his alchemical studies were spontaneous and free agents. The philosopher's mercury was "the subject of wonders" "a most puissant & invincible king," "the miracle of the world." The red spirit extracted from gold, Philalethes told him, "is ye vegetative soul & therefore makes ye barren dead bodies fructify exceedingly." When the quintessence is joined to gold and silver, such an excellent harmony is made "that it is impossible to express those infinite virtues which those excellent bodies acquire in that union and marriage."[119] Remorselessly Newton's measurements bound and confined that free spirit. In one paper he described the preparation of philosophers' aqua vitae.

> Observe that in destillation it is a lightsome & glorious spirit, being a Quintessence drawn out of the four first Elements & therefore not improperly called Spiritus Mundi. The matter of ten destillations gave about 50 1/20 Ounces Troy of Spirit besides the corrosive oyle, that is 4^{lib}, 2^{oz}, 1^{dwt}, of spirit.[120]

When he had finally constrained its activity into sharply defined categories, the alchemical spirit had become the force of modern science. For nearly thirty years Newton worked at alchemy. As it turned out, the primary subject of his alchemy was alchemy itself. From it he extracted the active seed which he planted in a new *terra alba*. I will not say that the feces he left behind was black and stinking, but certainly it was dead, and history in its progress could confidently cast it aside.

For Newton himself there was a final chapter. As I have indicated, the early 90's were a period of intense alchemical activity. Possibly it was the outcome of the *Principia*. An interruption the *Principia* may have been, but Newton had turned it to the advantage of his alchemy by building it on the concept of attraction. It had succeeded beyond his wildest dreams. One cannot remember too clearly how unplanned the *Principia* was. In August 1684, Newton had scarcely thought about mechanics for twenty years except for the brief exchange with Hooke in 1679-80. Halley's visit was an interruption, and it produced only a brief tract, *De motu*, devoted almost exclusively to the dynamics of orbital motion. Unexpectedly, however, the seed had fallen in fertile soil. The tract took on a life of its own, and as he worked out its implications, the concept of attraction led him on to the law of universal gravitation and the explication of a vast range of phenomena not even thought of in *De motu*. As he surveyed what he had achieved, he could not forget its further promise. He had explained the system of the world as far as its great motions were concerned, but there were innumerable other motions among the minute particles of bodies, such as the motions in fermenting, putrefying, and growing bodies. "If anyone shall have the good fortune to discover all of these, I might almost say that he will have laid bare the whole nature of bodies so far as the mechanical causes of things are concerned."[121] Here was a program adequate to explain his activity in the early 90's. Indeed, the six years following the *Principia*, what have appeared as a dormant period intellectually, culminating in the breakdown of 1693, suddenly take on new interest when viewed from the perspective of the art. Newton engaged in prodigious alchemical activity, both studying the record and experimenting. Nor was alchemy all. He also worked on a revision of the *Principia* and started to prepare the *Opticks* for publication. It was a period of mathematical activity. With Locke he exchanged his thoughts on Biblical criticism. I cannot resist the impression that Newton was in a state of spiritual exaltation. He had glimpsed the seductive outlines of Truth. Now, as Mars, he would embrace his Venus forever. His activity had even a political dimension. As he finished the *Principia,* the threat raised by James II had called him into the arena, and he had played an honorable role in the defeat of Popery and evil. In 1689, he found himself a member of Parliament and an object of William's patronage.

Perhaps most important of all, he had a disciple. "But no man can find ye work in his whole life," he had copied down among his *Notable Opinions* in alchemy, "without a Master."[122] The necessary counterpart of the master was the disciple, and in Fatio he now had one. The role of Fatio in Newton's life at this time is well known. I want to emphasize its alchemical dimension. Their correspondence is filled with the art. In his *Praxis,* Newton cited Fatio's letter of 4 May 1693.[123] Can Fatio have been the audience for whom those treatises were composed? And can Newton's collapse in the autumn of 1693, in which Fatio was clearly central, also have had an alchemical dimension? Was the collapse the inevitable conclusion of the state of exaltation when Truth in all her beauty refused to be finally possessed?

Such questions are, of course, like Newtonian queries, mere speculation. It is clear, nevertheless, that Newton's involvement in alchemy began to fade about that time. The cause was probably a great deal less dramatic. I have already suggested the possible role of the more orthodox scientific community in London. The new Warden of the Mint, who was manifestly pleased with the perquisites of the position, may have concluded that an occupation he had felt obliged to keep rather secret even in Cambridge was incompatible with his new dignity in the capital city. Perhaps there was also the growing realization that in fact he had embraced Truth, albeit in a somewhat homelier form than the enchanting Venus he had pursued. Whatever the cause, fade his involvement did, and that dramatically. A couple of his papers have notes from his early days at the Mint.[124] None of the papers seems to be much later in date. Newton's move to London was more than the conversion of a natural philosopher into a public servant. It was also the death of an alchemist.

Newton, a Sceptical Alchemist?

PAOLO CASINI

It may seem an act of presumption for someone who is not a Newtonian scholar but merely a curious amateur to comment on the work of such an expert as Professor Westfall. But what he has to say is of such importance that it will interest many people who are not professional historians of science, and will even fascinate historians of philosophy who instinctively react with diffidence to any mention of hermeticism and the world of magic.

In his recent writings, Professor Westfall has laid the groundwork for a reassessment of Newton's alchemical writings. His *Force in Newton's Physics* is the first history of mechanics from Galileo to Newton that deals seriously with "speculations" about the structure of matter that are generally excluded from orthodox histories of science. Westfall's careful analysis of Newton's manuscripts for the period that goes roughly from 1664 to 1684 enabled him to establish in a most convincing fashion that the development of key concepts such as inertia, force and gravity were not separated, in Newton's own mind, from complex "hypotheses" about the nature of matter. Newton's speculation concerning aether and elementary forces throughout the three decades from "An Hypothesis Explaining the Properties of Light" of 1675 to the first Latin edition of the *Opticks* in 1706, in what we know as Query 31, were largely influenced by the hermetic tradition and the naturalism of the Renaissance. This interpretation is developed and substantiated in Westfall's recent essay "Newton and the Hermetic Tradition"[1] and the paper in this volume.

Westfall has travelled further than anyone else into the no man's land of Newton's alchemy, and his exploration has yielded the following incontrovertible results. First, the alchemical manuscripts can be dated and correlated with Newton's *Principia* and his other published works. Secondly, the meaning and the sources of these manuscripts can be ascertained in spite of the obscurity of the idiom and the subject matter. Thirdly, Newton's abiding interest in the alchemical literature is strictly connected with his chemical experiments and his speculation on "active principles", on the cause of universal gravitation, and on the interaction between elementary corpuscles.

These results are the outcome of an open-minded approach to the problem of Newton's interest in alchemy. Westfall has avoided the pitfalls of both the old and the new historiography of science: he sees Newton neither as the rigorously positivistic scientist with nothing to hide, for whom "all is light", nor as the "last magician" or the "mystic" made fashionable by John Maynard Keynes. Instead of appealing to these stereotypes, Westfall constructs an image of Newton free of the constraint of post-Newtonian mental categories, an objective picture in the light of all the evidence, including that part which was discarded until recently. Alchemy was once considered the failing of a great genius, but, now that prejudices have been removed, the apprentice sorcerer of the manuscripts turns out to be not so strikingly different from the scientist who wrote the *Principia* and the *Opticks,* and who speculated on the nature of attractive forces and on the constitution of the aether.

In his previous studies, Westfall traced the connection from Newton's notebooks and his early notes on mechanics to his speculation on the nature of the aether in 1675. He described the intricate reasoning that led Newton to criticize the views expressed by Descartes, Gassendi, Charleton, and others. If atoms, the void and inert matter were characteristic of his mechanical philosophy, yet "Newton as a mechanical philosopher was acutely aware of the apparent presence of active principles in nature".[2] Hence, along with his acceptance of the mechanical philosophy, Newton felt the need of explaining a host of phenomena in terms of "plastic nature", and he turned to the *minima naturalia* of Henry More, the sympathies and antipathies of the hermetic tradition, and the active virtues of the Paracelsians and van Helmont. Without this joining of the mechanical philosophy and the hermetic tradition, we are at a loss to understand Newton's hypotheses about the aether and the role of "active principles". Nor can we understand the rise of the law of universal gravitation and Newton's concept of force. As Westfall points out: "The capacity of the mechanical philosophy to transpose any theory or conception into the idiom of material particles in motion allowed Hermetic ideas to operate within an overtly mechanistic outlook". This can be seen in Mayow, Digby, Lemery, even in Descartes and Boyle, the avowed critics of the hermetic tradition.

If I follow Westfall's line of reasoning correctly, he is in fact suggesting that the opposition between mechanical philosophy and naturalism, between atoms and active principles, is, in Bacon's phrase, an "*idolum theatri*" that we must get rid of. But if we grant this, how can we then say: "the fundamental question is the mutual interaction of the *two* in the development of Newton's scientific thought?"[3]

In any event, it is now clear that any "formal" reconstruction of Newton's dynamics will produce very little that is historically true. It is no longer possible to maintain that Newton discovered the inverse square law simply by following the mathematical way "and leaving aside, as many passages in the *Principia* state, the problem of the cause of universal gravitation. As a pure "mechanical philosopher", Newton would hever have reached his goal. It is wrong, therefore, to follow the received chronology and reason as though Newton had first formulated the law of universal gravitation for the planets and then extended the notion to explain chemical reactions, phenomena of cohesion, capillarity and the like. His highly involved notion of active principles—true centres of force within the elementary corpuscles of matter—enabled him to retain the "occult quality" of attraction banished by the Cartesians, to retain it, at least, as a natural phenomenon.

I need not stress the value of this viewpoint. Westfall's working hypothesis has the great advantage of allowing him to sketch a coherent picture (albeit one full of tensions) of Newton's various activities including his alchemical operations. This, after all, is the *experimentum crucis* of a good biography. Newton's alchemy is no longer seen as an aberration or a sick hobby but as an aspect of Newton's gigantic effort to introduce reason in every branch of human study. This is how I interpret Westfall's remark: "If alchemy influenced Newton, so Newton influenced alchemy". Can we not say the same of his sacred and secular chronologies, his biblical exegesis and his unitarian theology? Nourished in the "*prisca theologia*", familiar with the neoplatonic and hermetic tradition, Newton nevertheless turned the God of the Bible into a rational deity "very skilled in mechanics and geometry".

Newton's alchemical research was a rational attempt to conquer the irrational world of the occult. He was trying to repeat in his laboratory the mysterious synthesis of matter and force made by God when He created the world.

Westfall has found, if not the philosophical stone, surely Ariadne's thread to lead us out of the labyrinth of Newton's various theoretical and experimental interests. Nevertheless, I am left with one serious reservation. Is it correct to speak of "alchemy" in the full sense of that word? Doubtless, Newton manipulated alchemical formulas, but were his intentions alchemical? If we define them as such, are we really offering more than a metaphor? Westfall himself admits that "Alchemy was no one thing". How could it have been after the *Principia?* Could Newton have been pursuing the same ideal as the Renaissance magicians when his alchemical enthusiasm reached its peak in 1690? The cautious conclusions of Westfall suggest a dilemma that I shall try to express. On the one hand, Newton was sincerely convinced that the hermetic tradition contained a body of hidden truths that could be discovered through alchemical initiation and experiments in the laboratory. In this case, the mathematical physics of the *Principia* is but a small part of a much more ambitious dream: the discovery of all of nature's secrets. On the other hand, we could say that Newton, in spite of his alchemical activities, accepted with resignation the limits of the experimental method and rested content "to find a smoother pebble or a prettier shell than ordinary, wilst the great ocean of truth lay all undiscovered before me". In this second case, his alchemical research would have a much more modest significance and could be compared to an attempt to salvage what can be saved from a dilapidated intellectual heritage.

It is not easy to choose between these two alternatives. Perhaps Newton did not find it easy either. He may have oscillated between these views or he may, at various times in his life, have been assailed by doubts that we have no way of knowing. The long days that he spent transcribing and compiling alchemical recipes, his repeated efforts to extract the secret of attraction from the crucible and the alembic do not preclude that he may have felt himself to be—at least at certain periods—a sceptical alchemist. For instance, in a letter to Henry Oldenburg in which he discusses an experiment suggested by Boyle, he raises the doubt: ". . . if there should be any verity in ye Hermetic writers . . . whose judgmt (if there be any such). . .". The famous Query 31 of the *Opticks* not only rejects occult qualities but the very distinction between occult qualities and laws of nature rests on methodological principles that no alchemist would countenance.

> These principles I consider, not as occult qualities supposed to result from the specific forms of things, but as general laws of nature, by which the things themselves are

> formed, their truth appearing to us by phenomena, though their causes be not yet discovered . . . To tell us that every species of things is endowed with an occult specific quality by which it acts and produces manifest effects is to tell us nothing. . .

If we take these declarations seriously, then the "Rules of Philosophizing" and the "*hypotheses non fingo*" cannot be dismissed as sops for the non-initiated. Unless, of course, that Newton had a split personality—but I shall leave this conjecture to the Freudians. The temptation to clothe Dr. Newton in the garb of Dr. Faust is strong indeed. But to yield to this temptation without strong reservations would be to suppress the clear distinction between "occult qualities" and "the laws of nature", and ultimately between science and wild conjecture. The greatness and originality of the Newtonian synthesis would go by the board and post-Newtonian science would appear decadent and intellectually sterile.

Westfall's interpretation does not lead in this direction. He has made the facts available for us. The causes are and probably will remain unknown. But the scholar and the amateur (however sceptical and uninitiated) may yet be allowed to *feign* a few hypotheses.

Newton's Voyage in the Strange Seas of Alchemy

MARIE BOAS HALL

Professor Westfall has emulated Newton in the thoroughness with which he has examined Newton's alchemical manuscripts, and has thereby put us all in his debt. He has successfully established a chronology, (confirming the chronology apparent from Newton's experimental notebooks in the Portsmouth Collection) namely, that Newton's interest in alchemy deepened after his work in optics, and again after his completion of the *Principia,* but ceased altogether after his establishment in London. I feel sure that Professor Westfall is correct in his conviction that Newton's deep and prolonged reading of alchemical works led him to the conclusion that there was an underlying unity to alchemy, so that what different authors described under different names could be understood as applying to the same processes and substances, and that a comprehension of the alchemical "work" could be achieved by comparing and contrasting various terminologies in the manner of unravelling a complex code or cipher. Newton was very much a man of his own age, and it is worth noting that just as Newton compiled various versions of an *Index Chemicus,* so other, lesser men also sought to comprehend the mysterious language of alchemy in rational and quasi-rational terms. The last quarter of the seventeenth century saw the compilation and publication of a number of *Hermetic Dictionaries* (in English, Latin and French) which all purported to explicate the alchemical writings of the immediately preceding generation. We must all be most grateful to Professor Westfall for his patient, penetrating and painstaking reconstruction and study of Newton's alchemical writings.

To show the span of Newton's interest in alchemy, and his deep involvement in it, is a most useful effort, one which is of great interest to our present age, recoiling as it is from the pragmatism of an earlier generation and instead welcoming signs of irrational and mystic strains in that apparently most rational of human activities, the investigation of nature. I would not wish to underrate the importance of mysticism to Newton or to any other natural philosopher of the period: Robert Boyle, in many ways the most down-to-earth, least irrational of empirical natural philosophers of the 17th century, believed devoutly in the existence of an "incalescent mercury"—one that grew hot with gold—a "philosophic" (alchemical) substance that offered the key to the discovery of a method of transmutation. Boyle published two accounts of it, in 1675/6 (when Newton expressed doubt, especially of the wisdom of communicating such a partial discovery of Hermetic truth to the public) and again in 1678. This said, I find a certain difficulty in accepting Professor Westfall's thesis, that Newton was an alchemist *tout pur,* writing extensive treatises on alchemy, performing chemical experiments only for the purpose of comprehending it, and having, in general, a mystic end only in all his "labours with the furnace".

In other words, I question whether to show that Newton's interest in alchemy was deep, prolonged and evolutionary is to tell us all there is to know about this aspect of Newton's intellectual biography. Certainly among the alchemical MSS. we know, besides the very many which are mere reading notes—and Newton could not read even books in his own library without taking up a pen—there are other MSS. which read like Newton's *ipsissima verba.* No doubt they are; in one sense Newton certainly composed alchemical writings, that is, in the sense that he alone put together the words and sentences, even the ideas, in this particular form. But every text that I have ever examined at all closely—including those cited by Professor Westfall—seems to me to read like a summary of other men's ideas, and/or an attempt to corrolate and interpret such ideas. Newton compulsively wrote and rewrote everything he laid pen to, whether derivative or original. We have perhaps a dozen drafts of, say, the Preface to the *Principia;* it should not surprise us then to find half a dozen drafts of an attempt to understand alchemical treatises. Neither does it assist us in understanding what Newton was about. Professor Westfall has described the sociology of Newton's alchemical interest, if I may so express it; he has disclosed to us the full extent of Newton's attraction to mystic learning, and to a particular branch of it. But he has not attempted to explain what Newton did with his studies, nor what aims he had in mind. If Newton's alchemical interests are indeed worthy of study (and I believe with Professor Westfall that they are), then they are worth studying not merely descriptively and chronologically, but analytically, as we expect the *Principia* to be studied. It is more difficult, for to study the *Principia* is to study and understand success (on the whole, at least), and hence thought which has entered into modern science. To study Newton's alchemy and chemistry is to study failure, and a dead form of intellectual endeavour. But that does not mean that it should not be attempted.

Now to study the alchemical and chemical writings in the way I propose requires two different sorts of enquiry. The first is an investigation of Newton's sources; not who they were, but what they said, and how intelligibly they said it. The second is to examine Newton's conclusions, whether experimental or theoretical; not merely, that is, to perceive that he had certain ideas, but to try to discover what they were.

Let me take the question of sources first. The deep study of Newton's sources should embrace a perusal of English mediaeval alchemy, at least as enshrined in Ashmole's *Theatrum chemicum Britannicum,* followed by such early 17th century alchemists as Sendivogius, Basil Valentine, and the treatises in the 1659 *Theatrum chemicum;* next, later alchemists like George Starkey (who incidentally corresponded vigourously with Robert Boyle). Then it should also include the more chemical writers whose works were familiar to Newton: Boyle above all, but also lesser writers like David von der Becke. I think that if this were patiently done, the sources of many of Newton's statements would be more manifest, and we could better judge his originality. (To give an example: Newton's remarks about "Diana" and the form of trees must derive from the "tree of Diana" well known to 17th century alchemy and chemistry. It was described in the *Clavis Philosophicum* of Paul Eck de Sultzbach, first published in the *Theatrum Chemicum* of 1659, and is prepared by mixing aqua fortis (nitric acid) with mercury, and then adding this to silver dissolved in aqua fortis—all of which would certainly give odd but definite crystallisation effects. Moreover, Newton's quantitative care noted by Professor Westfall has *its* sources. These are partly chemical (Boyle's analytical efforts are relevant here), partly metallurgical (Humphrey Newton was presumably correct in recollecting that his master possessed Agricola, *On Metals*). Assaying had developed very refined techniques even in the 16th century, and if Newton possessed an assayer's balance, he could easily have weighed to fractions of a grain, for assayers commonly did so, though few chemists did, and no alchemist could.

Now if one is familiar with chemical as well as with alchemical writing of the 17th century, it becomes less easy than Professor Westfall appears to think to denominate all Newton's chemical writing as mystic, and to equate his description of processes with the belief in mystic forces of Nature. Where he sees mysticism and active principles, one can alternatively see chemistry and physics, even the mechanical philosophy. *Pace* Professor Westfall, I do not believe that any alchemist of any period except the present would see the "force" of *De gravitatione et aequipondio fluidorum* as mystic, any more than a 19th century spiritualist saw the aether of his physics as spiritual. It should be remembered that the force of the *Principia* or of magnetism could be regarded as an "active principle", but it could just as well be regarded as mechanical. One could argue equally that a man who had discovered that physical force could be mechanical and mathematical was more likely to think that alchemical principles could be so too than vice versa. As Newton said in the preface to the *Principia,* he had only succeeded for gravity, but he wished "we could derive the rest of the phenomena of Nature by the same kind of reasoning from mechanical principles. . ." [adding] "for I am induced by many reasons to suspect that they may depend upon certain forces. . . These forces being unknown, philosophers have hitherto attempted the search of Nature in vain. . .", but the *Principia* might be a model. Perhaps, therefore, Newton sought not a model, but phenomena in the study of alchemy.

But I digress. I think it is important to remember that the processes of chemistry have always attracted language more suited to the animate than to the inanimate. When I first learned chemistry in 1935, I was told that sodium and chlorine combined to form a compound difficult to dissociate into its elements again because "they wanted to fill up their electron shells exactly", since all atoms "wanted full electron shells". This is the language of alchemy, but I do not believe that my teachers were alchemists. Seventeenth century chemists talked about "fixed" and "volatile" salts without intending any biological connotation; they said a salt "runs *per deliquium*" where we say "it deliquesces", which is merely less picturesque. No modern chemist would find such expressions as "the menstruum wrought on the spelter continually till it had dissolved it" strange except linguistically; if he read "the solvent worked on the tin until it was all dissolved", he would know precisely what was meant. "Extract", "carry up" and so on can be accepted in supremely rationalist circles. To understand Newton's use of such terms, as to understand his use of analytical tests and his discussion of the nature of acids, one needs to have read 17th century chemistry, not merely 17th century alchemy. Only so can one properly judge, in my opinion, where Newton was being alchemical or mystic beyond the average. Even Newton's belief that he could understand alchemy was not unique to the age, for Boyle, confessing that he often could not understand Van Helmont at all, still thought he knew the method of preparing Helmong's *ens veneris,* and probably the wonderfully curative "Butler's stone". To my mind, more of Newton's thoughts on chemical processes transcend alchemy than Professor Westfall is prepared to admit. Perhaps significantly, for the Queries to *Opticks* Newton drew principally upon the drafts he had drawn up for conclusion or preface to the *Principia* in 1687, not upon his more strictly alchemical enquiries.

The second topic for investigation I proposed is that of Newton's conclusions, and here I think one must look particularly closely at the chemical notebooks, MSS. *Add 3975* and *3973*. These quite clearly describe chemical experiments, very precisely, even quantitatively expressed; and the fact that Newton talks about oaks, nets, the green lion, Diana, and so on no more *necessarily* makes them mystic than does the common use in contemporary chemical writing of astrological (or alchemical) names for metals. They spoke of Saturn or Venus; the English speaking world of the 20th century prefers to use "mercury" rather than "quicksilver" as a term. Similarly Newton's interest in volatilizing metals led him to develop low melting alloys — this is chemistry as well as alchemy. I believe that if one pursued Newton's alchemical writing sufficiently one might identify more of these mysterious entities than has yet been done. When this has been achieved then I think we may be in a position to understand what Newton thought he was trying to do. We can talk now about his interest in and attitude to alchemy, but we do not really understand it, because we do not yet understand his processes. When or if we do understand his processes, his purpose may surprise us.

Yet of one thing we may be sure: whatever his purpose, the results did not turn out as he expected. Somehow he hoped, as Professor Westfall has lucidly shown us, to penetrate behind the face alchemy presented to the world, to strip away the mystery and come face to face with the truth which, in touching tribute to his predecessors (whom he was always ready to name as giants), he firmly believed they had attained. When he did so, he only arrived in looking-glass land, where the truth that he sought vanished in optical illusion. Professor Westfall would have it that Newton confused illusion and reality, and accepted the alchemists' interpretations totally (at least for a time) rejecting the world of rational, mechanical, experimental philosophy. It may be that he is right. But I hope not; for I hope that the problem of Newton's motives and ideas is a genuinely interesting and potentially soluble one, which it cannot be if he was totally directed towards mysticism. If Newton was truly an alchemist, he was boringly pursuing a boring chimaera; and I find it difficult to believe that, whatever else he might be, Newton was ever dull. Much more plausibly intriguing is Professor Westfall's attempt (which he would, I am sure, agree needs further exploration) to link Newton's interest in the force of attraction with his interest in alchemy. I think that this is probably a just linkage. I would venture to suggest that Newton's earliest interests derive in part from his optical work, for colour had long been linked with chemistry, as indeed he linked it in 1675. Boyle's *Experiments and Considerations touching Colours* of 1664 begins as a work on optics and the mechanical philosophy (the nature of whiteness and blackness) but soon passes, after brief comments on the spectrum and the colours of soap bubbles, into chemistry and the development of colour tests for acids and alkalis. Certainly the suppressed *Conclusion* to the

Principia embodied many chemical ideas, which were to turn up again in *Opticks;* the antecedents of all of these have not yet been traced. The connections between Newton's alchemical researches and his practice of natural philosophy needs pursuing further in the manner ably begun by Professor Westfall, who has shown us that there is still a relatively untouched area of Newtonian studies.

Let me conclude on a minor point. This is that Newton's researches, whatever they may have been, were neither unknown nor frowned upon in his own day. Eighteenth century writers, Boerhaave and Pemberton among others, spoke frequently of his chemical experiments and chemical interests, deriving their opinions largely, but not wholly, from *Opticks.* The publication of *De natura acidorum* in John Harris' *Lexicon Technicum* in 1710 is also revealing. But let me remind you again of earlier manifestations. There is the "Hypothesis of Light and Colours" of 1675, not, it is true, published until 1756, but read to the Royal Society and by individual Fellows. Again, there is the letter to Boyle of 1678/9, published only in 1744, but probably known to others besides its recipient. Boyle and Newton both spoke of conversations on subjects of mutual interest. The friendliness, perhaps approbation, which Boyle felt for Newton is quaintly indicated in a scrap of paper preserved in the Boyle Papers in the archives of the Royal Society. It is interleaved into a MS. work entitled *Consilium philosophicum* dedicated to Boyle, by one "ER", and is inscribed in Boyle's hand: "Mr. Rothmaler's Booke yt I had from himselfe. by way of Gift as *I* understood it. Wch I bequeath to Mr Newton the mathematician of Cambridge". Clearly Newton never received it. Yet its existence among Boyle's papers, and the note by Boyle, suggest that Newton was not alone in voyaging in the strange seas of alchemy, although he travelled alone and went further—perhaps—than any man.

Hermeticism, Rationality and the Scientific Revolution

PAOLO ROSSI

I

In recent years discussions of the relations betwen *normal science* and *scientific revolutions* have attained a degree of refinement and sophistication reminiscent of late mediaeval scholasticism. Too often heated controversies concern the interpretation of a passage from a writer intent on commenting on another, who in turn was writing about a text by Galileo or Kepler, and it is difficult not to suspect that the meaning of the original has been irretrievably lost.

In many cases this suspicion is undoubtedly justified. Bacon or Galileo (and other writers as well) have become "symbols" for certain points of view, "masks" for inductivism or hypothetico-deductivism. The discussion of slogans such as "liberation from inductivism" or "the priority of hypothesis" replaces historical research itself. Historical figures are often reduced to "ideas incarnate", at the cost of a considerable impoverishment of our understanding of the historical process. Feyerabend's interpretation of Galileo's writings, (which, even in Italy, has been received enthusiastically) offers a good example. His description of Galileo as a scientist who systematically ignored those findings that contradicted his theories, and clung to his theories even when the evidence suggested he was wrong may be stimulating on a theoretical plane, but is hardly convincing historically. As P. K. Machamer has shown, this *mythical portrait* has been drawn by oversimplifying the problems involved and systematically distorting the texts.[1]

I am not sure to what extent J. E. McGuire is right in his attack on "the bloodless accounts of human thought that abound in the journals of academic philosophy, especially in the English speaking world". But I am sure he is right (possibly because I have myself said the same thing in Italy) when he asserts that historians of science "are not concerned with unalloyed conceptual nuggets, but rather with ramifications and interconnectedness of thought, be it in science, literature, politics and theology", and when he describes the historical process as an attempt "to catch the mind in action, not to fossilize timeless ideas".[2]

Faced with Feyerabend's *mythical Galileo* or the *legendary Bacon* of Popper and his followers, it is time that historians of science assert the need for a careful and detailed examination of the actual historical process. They must refuse to be reduced to the role of gatherers of exemplary cases to be used by philosophers of science as evidence for their theoretical constructs. I do not believe that the role of the historian of science should correspond to the one assigned by Bacon in his *New Atlantis* to the "Merchants of Life" and the "Mystery-men", while Agassi and Feyerabend fulfil the role of "Lamps".[3]

Some obvious facts are in danger of being forgotten in the course of these sophisticated discussions on the relations between normal science and scientific revolutions. The term *scientific revolution* (which roughly covers the period that goes from Copernicus to Newton) became common usage after Butterfield's lectures (1948) and the success of A. R. Hall's book (1954). As M. Boas Hall pointed out in her *Nature and Nature's Laws,* this term seems to have acquired greater credibility and legitimacy than a term such as "Renaissance", which has caused rivers of ink to flow, often with dubious results.[4]

Between the second half of the sixteenth and the end of the seventeenth century a new cosmology and a new astronomy came into being. It was the time of the first observations with the microscope and the telescope and of experiments on the vacuum. A new science of motion was born and the principle of inertia formulated, the circulation of the blood was discovered and a decisive step forward was taken in the study of anatomy and physiology. The theory of spontaneous generation was disproved and theories were formulated to explain how the earth was formed. A new distinction was drawn between the "subjective" world of everyday experience based on the senses and the "objective" reality of corpuscles moving according to definable laws.

Each of the innovations I have listed (with the exception of the last, which is usually included in histories of philosophy as well) would form an important chapter in any history of astronomy, physics, biology or geology. But we can only talk of "the scientific revolution" when we go beyond specialized histories of individual sciences and the arbitrary grouping of scientists into categories made to fit the subject matter of our university curriculum.

The legitimacy of the expression "methodological revolution of the seventeenth century" was not established either by specialized historians of single sciences, or by historians of science *exclusively* interested in the self-contained processes by which scientific ideas and theories are born of one another. Its legitimacy became obvious only once an idea of science had been accepted whose evolution was only relatively independent of the history of philosophical and religious ideas, as well as changes in social patterns.[5] This point of view stems from historical research accomplished along three main lines: intellectual history or history of ideas, sociology of science and science policy.

As I have mainly worked within the first tradition, I intend to stay within its limits, neglecting, quite deliberately, any reference to discussions on Puritanism by Weber or Merton. I believe that the period between the *De revolutionibus* and the *Principia* is quite rightly regarded as a turning point in the history of the world, not only because of important discoveries, new theories and novel experiments. It was a time when certain ideas and themes that are inextricably bound up with "science" came to the fore. These allow us to see the sudden break, the discontinuity that separates the new science from the old and helps us to understand some of the essential and decisive factors of what we usually call *modern thought.* I should like to list the following. First, the refutation of the priestly idea of knowledge inherent in hermeticism, in the alchemical literature, and in much of natural philosophy in the Renaissance. Second, a new appreciation of technical skills and mechanical arts, which as Leibnitz himself pointed out, led to consider the work carried out by 'empiricists' in their workshops and arsenals as a *kind of knowledge* that enriches man's true cultural heritage. Third, the new importance of scientific instruments as a means of making precise measurements and reproducing certain phenomena under controllable conditions. Fourth, the discovery of celestial bodies, plants, animals and men unknown before. Fifth, the birth, after a period of speculation on the existence of "savages" and the discovery of different cultures, of cultural relativism. Sixth, the idea of the plurality of other habitable planets and its consequence for the position and significance of man in a universe whose outer walls had crumbled. Seventh, the attempt, based on the new science of nature, to shape a new kind of ethics and politics. Eighth, the notion of the world as a machine not necessarily designed to suit man's standards, of a world therefore in which hierarchy had been abolished because all phenomena, like the component parts of a machine, have the same relation to the whole. Ninth, the conception of God as an engineer or watchmaker who follows a logical pattern in creating the world, but who does not intervene in its functioning. Tenth, the introduction of the dimension of time into the study of natural phenomena and the demise of the theory of an unchanging universe. Eleventh, the rise of the comparative anatomy that would lead to the "death of Adam". Twelfth, the notion of progress as the expansion of mankind and the idea of knowledge as the outcome of the efforts of several generations slowly accumulating results capable of integration and perfection. Thirteenth, the idea of collaboration and of publicizing the results of scientific research that lies at the heart of the first great scientific societies; and last of all, the theory that man can only know what he does or what he himself constructs.

The last topic on this list, so often neglected by those who study the *mechanisation of the world picture,* deserves a moment's attention. The affirmation that there is no substantial difference between the products of art and those of nature—supported by many seventeenth century thinkers—runs contrary to the Aristotelian definition of art as completing the work of nature or as imitating its products. Aristotelian philosophy (as well as Hippocratic medicine) sees nature as an *ideal* that art must imitate, indeed as a *norm* whose precepts art must follow. Should art assert that it can equal the perfection of nature then it is treated—by the medieval doctrine of *imitatio naturae*—as an example of ungodliness: art is trying to counterfeit nature. The mechanical arts are necessarily "*adulterinae*" because they must borrow their movement from nature. According to Francis Bacon, this doctrine is connected with the Aristotelian theory of species, according to which a product of nature (e.g. a tree) has a *primary form,* whereas a product of art (e.g. a table) has only a *secondary form.* This is why Bacon's projected *History of the Arts* which was to complete the *Natural History* is so important:

> And I am the more induced to set down the History of the Arts as a species of Natural History, because an opinion has long been prevalent, that art is something different from nature, and things artificial different from things natural; whence this evil has arisen, that most writers of Natural History think they have done enough when they have given an account of animals or plants or minerals, omitting all mention of the experiments of mechanical arts. But there is likewise

> another and more subtle error which has crept into the human mind; namely that of considering art as merely an assistant to nature, having the power indeed to finish what nature has begun, to correct her when lapsing into error or to set her free when in bondage but by no means to change, transmute, or fundamentally alter nature. And this has bred a premature despair in human enterprise. Whereas men ought on the contrary to be surely persuaded of this; that the artificial does not differ from the natural in form or essence, but only in the efficient.[6]

Within the mechanistic picture of the world, a machine, whether real or merely imagined, functions as an explicative model. It becomes an adequate representation of reality based on quantitatively measurable data, in which each element fulfils a function that is dependent on the configuration and the motion of the whole. To know reality means to understand how the world-machine functions. And a machine can always, at least in theory, be taken apart and put together again. "We enquire into things in nature," Gassendi writes,

> in the same way as we enquire into those things of which we are ourselves the authors. . . . Wherever possible in the study of nature we make use of anatomy, chemistry and other aids so as to understand, by breaking down the bodies as far as possible and dividing them as it were into their component parts, what these elements are and what manner of criteria helped in their composition, and to see whether, by following different criteria, they would have or still could become other than what they are.[7]

The end of Gassendi's statement is particularly significant. The world of phenomena that can be reconstructed with the aid of scientific analysis, and the world of artificial products, which have been constructed or reconstructed intellectually or manually, are *the only realities of which we are able to have true knowledge:* the new science is interested neither in *"quidditates rerum intimas"*, nor in *"arcana naturae"*: rather it is based on a phenomenological knowledge of the world.[8]

We can only have full knowledge of machines (the artificial products of man) and of what can be interpreted mechanically. Several basic principles of Aristotle's notion of the relationship between art and nature, which had dominated the whole of European culture, were thus deliberately rejected. Descartes, like Bacon, draws no distinction between natural and artificial objects. Lightning, which according to the ancients could not be copied, has, in fact, been imitated in modern times. Art does not ape nature, nor does it "kneel before nature" as a popular medieval tradition asserted. Descartes is explicit on this point: "There is no difference between the machines constructed by artisans and the divers bodies that nature composes". The only difference is that the mechanisms of man-made machines are visible, whereas "the tubes and springs of natural objects are often too small to be perceived by the human senses".[9]

The Platonic idea of God as a geometer was modified by the idea of God as a mechanic, artisan of this perfect clock that is the world. Knowledge of ultimate causes and essences, which is denied to man, is the prerogative of God as creator and *constructor* of the world-machine. The criterion of *knowledge-as-making* or the *identity of knowledge and construction (or reconstruction)* holds therefore not only in the case of man, but of God also. The human intellect is finite and limited, it can apprehend only those truths that have been constructed by man. We are really able to know only that which we make ourselves, or rather that which is *artificial.*

Broadly speaking, in so far as nature is not conceived of as an artifact, it is unknown and unknowable.

"Il est difficile", Mersenne writes in his *Harmonie Universelle,*

> *de rencontrer des principes, ou des veritez dans la Physique, dont l'objet apartenant aux choses que Dieu a crées, il ne faut pas s'estonner si nous n'en pouvons trouver les vraies raisons, et la maniere dont elles agissent et patissent, puisque nous ne sçavons les vraies raisons que des choses que nous pouvons faire de la main, ou de l'esprit; et que de toutes les choses que Dieu a faites, nous n'en pouvons faire aucune, quelque subtilité et effort que nous y apportions, joint qu'il les a pû autrement faire.*[10]

Although Hobbes maintained very different views from those of Mersenne, his conclusions are not dissimilar on this point:

> Of arts, some are demonstrable, others indemonstrable; and demonstrable are those the construction of the subject whereof is in the power of the artist himself, who, in his demonstration, does no more but deduce the consequences of his own operation. The reason whereof is this, that the science of every subject is derived from a precognition of the causes, generation, and construction of the same; and consequently where the causes are known, there is place for demonstration, but not where the causes are to seek for. Geometry therefore is demonstrable, for the lines and figures from which we reason are drawn and described by ourselves; and civil philosophy is demonstrable, because we make the commonwealth ourselves. But because of natural bodies we know not the construction, but seek it from the effects, there lies no demonstration of what the causes be we seek for, but only of what they may be.[11]

The idea of *knowledge* as *construction* and *making* was bound to have a profound influence on ethical and historical thought, and this passage from Hobbes has been compared to Vico's principle of *verum-factum.* "We demonstrate geometric propositions because we make them; if this were possible we could do the same for physics". In experimental physics, as he pointed out in *De Antiquissima Italorum Sapien-*

tia, "we hold something true in nature only when we can make something similar, by means of experiment", and "therefore arithmetic and geometry, as well as mechanics which depends on them, are within the reach of man's faculties, since in all three we prove a truth only in so far as we construct it". Hence the comparison between what man makes and what God makes: "Just as nature imparts life to physical things, so the human mind imparts life to mechanics; just as God is the artificer of nature, so is man in things made by art". In the *New Science,* the world of history is conceived of as having been made and constructed by man:

> But in the night of thick darkness enveloping the earliest antiquity, so remote from ourselves, there shines the eternal and never failing light of a truth beyond all question: that the world of civil society has certainly been made by men, and that its principles are therefore to be found within the modifications of our own human mind. Whoever reflects on this cannot but marvel that the philosophers should have bent all their energies to the study of the world of nature, which, since God made it, He alone knows; and that they should have neglected the study of the world of nations, or civil world, which, since men had made it, men could come to know.[12]

After the second half of the sixteenth century, philosophers and scientists began to look upon 'experience' in a different light.[13] As manual labour and the mechanical arts acquired a new status and a new importance within the encyclopaedia of knowledge, a new understanding of the relations between *knowing* and *doing* was born. Once the identification of *knowing* and *doing* had forced researchers to give up the idea of ever understanding the "essential" structure of nature, it was bound to influence ethics, politics and history, with consequences that can hardly be underrated.

II

The intellectual stance of magicians, alchemists, Paracelsians, and hermeticists played a not indifferent role in the gradual acceptance of the new way of considering "experience" and "doing".

The so-called "hermetic tradition" has become the subject of a learned controversy among English-speaking historians of science after the publication of F. A. Yates' important book on Giordano Bruno in 1964. The importance of magic and astrology in the history of modern thought, the historical dimension of the Renaissance taste for magic, and the profound difference between the mediaeval condemnation of magic and its praise during the sixteenth century, had been emphasized as early as 1950 by Eugenio Garin in two important essays: *"Considerazioni sulla Magia"* and *"Magia e Astrologia nel Pensiero del Rinascimento"*, both republished in 1954 in *Medioevo e Rinascimento.*

I dedicated a large portion of my book on Francis Bacon (first published in Italian in 1957 and in English translation in 1968) to illustrate Bacon's profound indebtedness to the magico-alchemical tradition and to the Renaissance concept of magic. Significant contributions to these problems were made by Walter Pagel (whose important work dates from the 50's) and later by D. P. Walker, A. G. Debus, J. E. McGuire, P. M. Rattansi and M. C. Jacob. After the enormous bulk of research on this subject it is no longer possible to sport a 'positivistic' attitude, or to claim to be totally uninterested in aspects of culture that, at least during the first half of the seventeenth century, were not considered the residue of dark mediaeval superstitions, but the ancient and respectable heritage of ideas to which not only Bruno, Campanella, and Robert Fludd helped themselves generously, but also such thinkers as Bacon, Gilbert, Kepler and Newton.

When I prepared a new edition of my book earlier this year, I kept the subtitle *From Magic to Science.* But since I find myself somewhat at variance with F. A. Yates and P. M. Rattansi, who tend to see Bacon as an exponent, in a more modern idiom, of the ideas and values of the hermetic tradition, I should now stress Bacon's portrait of "the man of science" found in so many parts of his work.[14] As years go by, I am more and more convinced that to explain the genesis—which is not only complicated but often confused—of some modern ideas is *quite different* from believing that one can offer a complete explanation of these ideas by describing their genesis.

In the 70's, the image of Bacon as the "father" or "founder" of modern science is giving way to that of Bacon as the "transformer of hermetic dreams". In contrast to what was happening in the 50's when I started to work on Bacon it is now dashionable to emphasize the importance of Orpheus, Zoroaster, Hermes and the themes of *ancient theology* in the philosophical and scientific writings of the seventeenth century. What started off as a useful corrective to the conception of the history of science as a triumphant progress, is becoming a retrospective form of historiography, interested only in the elements of continuity and the influence of traditional ideas. Bacon and Copernicus, Descartes and Newton are considered in their links with the past, and in their common 'descent' from earlier revolutions and cultural innovations. We are in danger of ignoring, as totally irrelevant, the ideas, theories and doctrines that make these writers difficult to fit into the endless list of writers on hermetic subjects or of exponents of rhetoric who published their works between the second half of the sixteenth century and the end of the seventeenth century. Instead of asking what new ideas have given these giants their place in history, too often historians consider exclusively what existed in the past or may be derived from the past without unwanted residues. This abuse of the category of "persistence", and the tendency to indulge in retrospective history, as well as the insistence on the unity and continuity of European culture from Petrarch's *Secretum* to

Rousseau's *Contrat Social,* and the "Warburghian" taste for symbols and magic may lead to conclusions as prejudiced as those that resulted from the use of the idealistic categories of "anticipation" and "caducity".

Bacon, who was extremely fond of classifying and typifying, saw in magic a typical manifestation of *phantastical learning,* in scholastic disputations a kind of *contentious learning,* and in Ciceronian humanism the expression of *delicate learning.* The fact that he was variously conditioned by these three cultural forms should not blind us to the fact that he attempted to formulate a new image of science in violent contrast with magic, scholastic thought and the tradition of humanism. Indeed in the latter he saw the expression of a "*fucata et mollis*" philosophy that could serve civic ends but was limited to stylistic arabesques and verbal solutions and ultimately dangerous to the "austere search for truth".

In stressing the hermetic themes of the interconnection between theory and practice, and the unity of the products of art and nature, or in returning to the idea of man as the servant-lord of nature, Bacon was making a new use of fourteenth century models. He altered the ideas he borrowed and introduced them into a novel context. For him the separation of science from theology is basic, and Platonism is "detestable" because it reduces natural phenomena to spiritual precepts in a hierarchic and "ascendant" world-picture. Hence also Bacon's distaste for those "moderns" who hoped to found a system of natural philosophy on the book of *Genesis* or on other parts of the Scriptures. As White has shown, the *New Atlantis* can hardly be accused of avoiding the subjects of exemplarism and symbolism,[15] but we should not forget that, in rejecting a whole world-picture, together with the doctrine of man as a microcosm of the universe, Bacon also did away with the idea of the world as the "living image" of God. God "only resembles Himself, quite apart from any metaphor": a study of material things cannot throw any light on his nature and will. With regard to divine mysteries science is dumb. To speak of Bacon's religion is to speak of his physics: if an examination of the world can reveal nothing about God, if reading the book of Nature must be held strictly separate from reading the Scriptures, then the discovery and analysis of forms, latent processes, "*schematismi*" and "*metaschematismi*", cannot reveal any kind of divine presence, nor any creative power at work in the world. When he asks men to humbly leaf through the book of living creatures and to give up all attempts to build the ship of philosophy out of a rowlock or a shell, or to direct their energies to the writing of a great history of nature and the arts, Bacon, then in his sixties, mentions Giordano Bruno for the first and only time.[16] In the same breath he recalls Patrizi, Telesius, Pietro Severino, Gilbert and Campanella, as philosophers who arbitrarily created the subjects of their worlds, as if they were dealing with fables following one upon another on a stage. According to Bacon, man is not at the centre of secret correspondences; the universe is not a web of symbols that correspond to divine archetypes; scientific research in no way resembles an incommunicable and mystical experience.

III

It is well known that in defending the central position of the Sun, Copernicus calls upon Hermes Trismegistus and reveals the influence of the magico-hermetic world-picture as it had been elaborated by Marsilio Ficino. William Gilbert also refers to Hermes and Zoroaster when he links his doctrine of magnetism with universal animation. Bacon talks of the "perceptiveness", the "desires" and "aversions" of matter; and he is generally influenced by the language and the models prevalent in the alchemical tradition. It is from the world of magic that he derives his definition of man as the servant and interpreter of nature which replaces the venerable definition of man as "a rational animal". Kepler displays a profound knowledge of the *Corpus hermeticum,* and his conviction that there is a secret correspondence between the structures of geometry and those of the universe, as well as his theory of a celestial music of the spheres, is imbued with Pythagorean mysticism. Tycho Brahe persists in considering astrology as the legitimate and practical application of his science: "I am used," he writes in 1589, "to calling alchemy the terrestrial astronomy since the subjects under examination bear analogy to the celestial bodies and to their influence." Galileo himself, always so clear and rigorous, and in whose writings there are no concessions to mysticism or magic, mentions Dionysius the Areopagite in a letter and speaks of "an extremely spiritual substance that warms, gives new life to all living creatures and makes them fecund". Descartes, who in his maturity rejected all forms of symbolism and whose philosophy became a by-word for rational clarity, placed, in his youth, the works of the imagination before those of reason. Like so many magicians in the sixteenth century, he enjoyed constructing automatons and "shadow gardens"; like so many followers of Raymond Lull he insisted on the unity and harmony of the cosmos: "There is one active force, love, charity, harmony in things. . . . Every corporeal form works by harmony". These themes reappear in a different key in Leibniz, whose logic blends theories taken from the Lullian tradition of hermeticism and cabalism and the dreams of Comenius' *pansophia.* His "*scientia generalis*" not only embraces logic, it is also an "*ars inveniendi*" and a "*methodus disponendi*", it is synthesis and analysis, didactics and the science of teaching, the art of remembering or mnemonics, "*ars characteristica*" or symbolic art, philosophical grammar, Lullian art, the cabala of sages and natural magic. Leibniz was convinced that it was possible to establish a general science that would discover the full correspondences between the original forms that make up reality. One might add that his idea of harmony is founded on texts that one can hardly call "scientific". In the *De Motu Cordis,* William Harvey's praise of a unique and central government of life where the heart plays the role of the Sun in the microcosm is compounded of heliocentric astronomy and cardiocentric physiology. Even the Newtonian concept of space as the *sensorium Dei* can be traced to Neo-platonic influences and the Judaic Cabala. Recent studies of Newton's manuscripts reveal his faith in an *ancient theology* (a central theme in hermeticism) whose truth had to be established with the aid of the new experimental science.[17]

Nevertheless, the influence of the Renaissance tradition of magical naturalism, the wide diffusion of books on magic, alchemy and astrology, the prestige of astrologers and the persistence of hermeticism well into the eighteenth century do not alter the fact that themes taken from the hermetic tradition were incorporated into a different framework, and were moulded to different ends. It is hardly accidental that the attack on the obscurities of magic, the pretentious illusions of the alchemists, the deceptions of astrology, are to be found in all those writers who, for various reasons, may be numbered among the champions of the scientific revolution. For Kepler, the fundamental characteristic of magical thought is the attempt to describe and understand nature by using 'hieroglyphics' that cannot be rationally discussed. He opposed those who enjoyed "the shadowy enigmas of things" and sought to "subject to the clarity of the intellect what is wrapped in darkness". The clarity of mathematical analysis was the reverse of the obscure and allusive language of magic. The scornful silence of Galileo and Descartes, Bacon's aggressive attack, Gassendi's opposition to Fludd, Mersenne's long struggle against the practitioners of the occult, Boyle's ironic comments on Paracelsus' followers, Pierre Bayle's invective against the "shameful superstitions" of astrology are revealing. Different people from varying points of view all cry out against a mystical world picture and appeal for greater linguistic clarity, for models that can be checked and experiments that can be repeated.

These are the reasons, rather than the vague formulation of the inductive method, why so many, from the first decades of the seventeenth to the beginning of the nineteenth century, held Bacon's writings to mark the beginning of a new era. In science, effective results, as opposed to the illusory claims of the magicians and alchemists, can be reached only by the cumulative results and the cooperation of scientists. Bacon waged war against the magico-hermetic ideal of the sage as initiated and "inspired". He denounced knowledge that leads to results that cannot be transmitted to the layman and condemned an attitude of mind dominated by the category of the marvelous. He insisted on the organizational and institutional aspects of science because he was convinced that the end of scientific knowledge was not to be confused with personal success, nor with the desire to excite popular admiration for the sage.

This image of science was generally accepted by the first practitioners of modern science. The rigour of logic, the publication of results, the desire for clarity were championed energetically within the ambit of a culture that did not accept them as obvious and in which people continued to hold beliefs and attitudes and ways of seeing the world that appeared to offer *a real cultural alternative* to the new learning and the new methods.

As often happens, the historical process reveals elements of continuity and of discontinuity, the slow transformation of the meaning of words and expressions and the laboured acceptance of new ideas within a traditional context. Far too often we tend to forget that the world of the natural philosophers of the period that goes from Copernicus to Newton was not the world that had seen the emergence of the figure, the outlook and the social function of the modern scientist. In this world, the methods, the experiments, and the institutions of modern science were a thing of the future.

After the methodological revolution of the seventeenth century the first specifically modern attitudes to the world were born. They were still linked with the past and with the new interpretation of Archimedes, Galen and Pappus. Often they appeared to be mere revivals and rediscoveries of ancient themes, yet they were qualitatively different from those that had been characteristic or dominant in the ancient world, the Middle Ages or the Renaissance.

During that period the image of science took shape and was, for good or ill, to have a decisive effect on subsequent history. The new *image of the man of science* elaborated at that time was different from those of the ancient philosopher, the saint, the monk, the university professor, the courtier, the perfect prince, the artisan, the magician, or the humanist. The aims of the composite intellectual groups that contributed to the development of scientific knowledge during the second half of the sixteenth century and during the seventeenth century, and the values they upheld, were certainly very different, as well as more "impersonal" than those of individual sanctity or literary fame.

Within the divergent schools, trends, and attitudes typical of the age of the scientific revolution, the complex phenomenon we call "science" developed with its categories, its ways of thinking, its choices of value, its methods and its institutions. These categories, methods and institutions were intimately connected with a body of speculations about the nature of science, the different ways of approaching reality, the relations of theory to experiment, the value of hypotheses, the responsibilities of scientific knowledge, its historical function, its limitation and its ties with religion.

The image of science that came into being in that period already had something that has become characteristic of modern science: learning conceived as a perfectable construction built upon cooperation and requiring a specific language and specific institutions. A knowledge that can develop and aims at "true" statements which must be subjected to the test of experience and comparison with alternative theories.

IV

If I disagree with F. A. Yates and P. M. Rattansi's conclusions, it is not because they point out the necessity of studying magic and hermeticism, but because they emphasize only the elements of continuity between the hermetic tradition and modern science. I believe that Rattansi has perfectly good reasons for rejecting the rigid "canons of rationality" of M. B. Hesse who seeks "deep reasons" that can be singled out "independently of their historical incarnations".[18] I agree with Rattansi that it is quite futile to try to rid oneself of the problem of the interconnections of the hermetic tradition and modern science by labelling hermeticism "a pathological resurgence of irrationalism". Epistemological relativism does not loom in every attempt to study the social context of scientific knowledge, nor should one interpret every attempted study of hermeticism and Neo-platonism as a plea for a mystical approach to nature. Rattansi is right in stressing that the study of nature is related to larger metaphysical assumptions and is involved in complex ways with other areas of intellectual culture, and that the Neo-platonic cosmology is important for an understanding of Copernicus' arguments in favour of heliocentrism. Likewise the context of ideas which is relevant for understanding some of Newton's fundamental scientific ideas is much broader than is usually realized. I share Rattansi's conviction that our main task as historians of science is not to show "the timeless rationality of Newton's scientific inferences", but to research into "the complex and changing interplay between Newton's scientific concern and a whole variety of other concerns, and between them and the society and intellectual culture of his own time".[19]

Beyond these points of general agreement, however, I am ill at ease concerning some of the conclusions reached by Rattansi and Yates. For what exactly does Rattansi mean when he refers to "our own sort of science" or "our own sort of rationality"?[20] Is research into the fields of hermeticism and magic important because it helps us to gain a clearer understanding of the origins of modern science (which indubitably drew strength at the beginning of its long life from a "disreputable structure of ideas") or it is important because it leads us to conclude that modern science is merely the continuation, in a new guise, of the mystical approach to nature? Is it important because it shows us how tortuous and complicated the progress of scientific reason is or because it reveals the irrational foundations of scientific knowledge? To use a more fashionable terminology, do hermeticism and modern science follow one upon the other like two metaphysical research programs that are basically equivalent? Are they two incommensurable paradigms each of which contains its own standard of rationality? Within the broader sphere of the history of mankind, did the scientific revolution imply a new kind of knowledge that is both inter-subjective and capable of growth? Are we questioning our inadequate understanding of the genesis of modern science or modern science itself?

These queries would appear unjustified if in the last few years, in a far wider sphere than that of the relation between the hermetic tradition and modern science, irrational views had not become the starting point of an analysis of science. Thomas Kuhn declares he does not understand what his critics mean "when they employ terms like 'irrationality' to characterize my views".[21] I fear that his bewilderment may be solely dependent on his limited acquaintance with the classics of contemporary "philosophy of life".

Kuhn has tried to dissociate himself from the consequences of Feyerabend's points of view. But beyond these attempts at self-defence, it still holds true that he conceives the transformations of science, the passage from one paradigm to another, as a kind of mystical crisis and a religious conversion that may be examined only in the light of a *psychology of discovery.* His history of science reads like the intimate diary of a follower of a mystical philosophy. Between intervals of stasis or *normal science,* new paradigms emerge each of which is incommensurable with the preceding ones. "Each paradigm", as Lakatos pointed out, "contains [according to Kuhn] its own standards. The crisis sweeps away not only the old theories and rules, but also the standards which made us respect them. The new paradigm brings a totally new rationality. There are no super-paradigmatic standards. The changing is a bandwagon effect. Thus in Kuhn's view scientific revolution is irrational, a matter for mob psychology".[22]

These implications were carried to their logical extreme by Feyerabend whose philosophy is dominated by appeals to "faith", "choices", the incomparability of paradigms, the incommensurability of theories, and the *Gestalt-switch theory.* All of which he uses to illustrate the transformations and mutations that have occurred in the history of science. When asked whether science is irrational, Feyerabend answers with a string of *yes* and *noes* and seems quite content to assert that in scientific knowledge there exist "ingredients which defy a rational analysis". The ultimate scope of his argument, as may be seen quite clearly in his conclusions, is to show the radical discontinuity and substantial irrationality of the scientific venture. His writings show the influence of all the basic themes, so fashionable today, of the Neo-Romantic revolt against science at the turn of the century. The contrast between positive subjectivity and mere objectivity, the parallel between science and poetry, the choice of theory reduced to a "matter of taste". He does not even omit the catchword of contemporary *counter-culture:* "we may construct a world in which science plays no role whatever: such a world, I venture to suggest, would be more pleasant than the world we live in today".[23]

This combination of irrationalism and the exaltation of the primitive and the untainted is one of the traditional themes of the *Zerstörung der Vernunft.* But the idea of describing the choice between rationalism and a Heideggerian attitude as one between a "dragon" and a "pussy cat" (in which Heideggeranism is the pussy cat) had not occurred to any one else before.

However, let us return to something that is closer to my subject. The consequences of these theories have been pointed out by Machamer:

> Why should anyone prefer a New Revolutionary Theory to an Old Accepted Theory. . .? To this Feyerabend suggests a reply which has all the hallmarks of a religious faith. Galileo through some leap of "genius" had faith in Copernicanism in the teeth of all evidence, and in spite of all that is rational. Such faith is no doubt allied to the more traditional view concerning the genius of discovery. In the light of such faith Galileo propagandized and proselytized on behalf of the new theory. There can be no question as to how the new theory might be an improvement on the old one, how it provided new and better explanations.[24]

I suspect that a history of science inspired by the methodological canons of Kuhn and Feyerabend will enjoy a considerable success in Italy as well as elsewhere. If only the world-pictures or the general conceptions of reality are important in the history of science, then it can be entirely subsumed under the history of philosophy. We may forget, as useless digressions, research into the structure of theories which require specialized training and research into the sociology of learning and the history of scientific institutions. Burtt and Koyré, Kuhn and Feyerabend will be instrumental in confirming an *Old Accepted Theory,* still much alive in Italian culture. They will be used to show the truth of the view, propounded at the turn of the century by Giovanni Gentile, that science does not exist and is ultimately absorbed into philosophy.[25]

V

Against a history of science that overrated experiment and underrated the role of hypothesis and theory, it was necessary to stress the importance of "intellectual courage" and "speculative audacity". It was right to reject a history of science that described its development in a triumphant key, as a linear process of progressive growth and to emphasize the tortuous, non-linear and by no means inevitable nature of the historical process of science. It was important to decry an autonomous and "separate" conception of the history of science and to emphasize the presence of philosophical ideas, and religious myths in the development of science.

But things did not stop here. The destitution of the positivistic, Baconian idea of an inductive science has been made to coincide with a crisis involving the entire rationality of science. The theory that scientific thought does not develop along separate lines has been made to coincide with a denial of the relative autonomy of scientific theory. The theory that the histories of scientific thought and philosophy cannot be separated has been identified with the theory that the history of science is nothing more than the history of world-pictures. The theory that scientific theses are conditioned by extraneous elements to the thought processes (for example, a given image of science which mirrors the culture of a given period) is now taken to mean that a rational analysis of theories is no longer possible.

The recognition of the existence of a whole series of problems and difficulties has been transformed into a string of impossibilities: *since* experiment no longer plays the decisive function attributed to it by inductivism, *then* science contains only theories; *since* a linear and continuous development is absent from science, *then* only a series of "choices" between theories remain; *since* any attempt at a reconstruction on rational lines stumbles on elements that cannot be reconstructed, *then* one is left with appeals to individual or collective psychology.

The history of mankind is full of myths, of religious and metaphysical ideas, of conjectures, of daring hypotheses, and of modes of perceiving and understanding the world that will never be integrated into science. In other cases, however, some of these myths, hypotheses, conjectures, influence science and become an integral and component part of science: *they "dissolve" into science. They become difficult to recognise and operate by stealth within science, in such a way as to defy philosophical interpretation and attempts to explain their presence.*

Should a recognition of these "hidden presences" allow us to reduce the whole structure of scientific knowledge to these elements and to dissolve it into them? Biography and psychological history may be able to reveal the motives and unconscious roots of the creative processes of thought. An understanding of these hidden presences can help us to understand Newton, they can also serve to throw light on certain aspects of his scientific thought, and they can help us to see some aspects that we might have missed or underrated. Seen in this light Frank Manuel's work is certainly stimulating.[26] But once we have described the Newtonian concept of absolute space as his personal answer to the threat of cosmic anxiety, the inescapable fact remains that this concept is made up of mathematical, physical, theological and philosophical elements that can hardly be reduced to psychology or entirely explained by it. It appears important to recognize the variety of levels and planes in all forms of historiography. Many men have felt weak and insecure in childhood, many have been imbued with a sense of predestined mission, many have identified God with the "absent father" figure without building a system of natural philosophy. Similarly, many twentieth-century writers had phobias and anxieties, but only Svevo sat down and wrote *The Conscience of Zeno*. Explanations of Newton's and Svevo's neuroses may well be part of the story, but they are not the whole story. The "cultural objects" created by Newton and Svevo do not appear in history exclusively as products of their neuroses.

The recognition of the "hidden presences" within the hermetic tradition of modern science does not entitle us to *reduce the latter to the former,* and to forget that in the case of the history of science—at least from the age of Galileo and quite apart from what was happening in the world of magic—it is justifiable to speak of theories that are more or less rigourous, have greater or less explanatory and/or predictive power, and are verifiable to a greater or lesser degree.

Lakatos was quite clear on one point: the controversy between Popper's "rationalism" and Kuhn's "irrationalism" does not refer to technical problems of epistemology: "it concerns our central intellectual values, and has implications not only for theoretical physics, but also for the underdeveloped social sciences and even for moral and political philosophy. If even in science there is no other way of judging a theory but by assessing the number, faith and vocal energy of its supporters, then this must be even more so in social sciences: truth lies in power",[27] or, one might add, it belongs to those who possess the best techniques of persuasion.

In insisting on the *comparison of frameworks,* and the possibility of rational as well as psychological analysis, one should also make clear that this capacity for comparison and analysis does not go to make up the essence of man, nor is it to be considered an eternal category of the spirit. *Rather they are tied up with the history of the modern scientific mind,* with its dissociation from the world of magic, and its rejection of the tradition that had perpetuated, in an extremely refined guise, the hermetic attitude of mind well into the seventeenth century.

When we speak of a revolution with regard to the birth of Baconian and Galilean science, I believe we should underline the emergence, as in all other revolutions, of something new that cannot be entirely explained by what was before. Why neglect the obvious? This revolution gave rise to an historical process that has transformed the world in quite a different way from either religion or philosophy. Science and technical skills are responsible for the first attempts at a cultural unity of the world. Some methods of knowing and controlling nature tend, from the moment of their inception, to become the common property of mankind. Discussion of the ends of science should be kept separate from a discussion of the scientific world-picture. It is certainly true, as Ravetz has pointed out, that to avoid the "tragedy" of modern science "something more will be required than the very straightforward intellectual integrity which has sufficed as the basis for the ethic of science in simpler times".[28] But it is also true that the problem, which is mainly political, of the *uses of science* cannot be reduced to appeals to a freedom that could be reintroduced into history after the human intellect has been destroyed. Doubtless the study of the interconnections between hermeticism and modern science have greatly enlarged our historical horizon. But the history of science is not alone in experiencing enlargements of this kind. A knowledge of the world of magic has had a decisive influence on cultural anthropology as well. Some cultural anthropologists, as well as some historians of science have accepted the absolute *equivalence* not only of all forms of culture, but of all possible world-pictures as well, and have put forward primitive and irrational theories. It is one thing, however, to become aware of the origins and the historical limitations of our own civilisation, and another to renounce our intellectual faculties, praise Shamanism, and cultivate the occult and the prophetic as new and superior forms of knowledge.

In the so called "literature of dissent" enthusiasm for the new magic is rampant. In *The Making of a Counter Culture: Reflections on the Technocratic Society and Its Youthful Opposition,* Theodor Roszak appeals to the teaching of Thomas Kuhn to reduce "objective conscience" to "mythology". Roszak contrasts the extraordinary possibilities of a new magical world-picture with the narrow rationality of science, since science "blunts our sense of the marvelous" while Shamanism offers a new, freer culture.

This is not an isolated case: a return to the archaic phase of magical experience is accepted by many as a valid method of freeing ourselves from the sins of our civilisation. But this cheap liberation is an illusory escape into the past, mere nostalgia for a golden age, a convenient flight from historical responsibilities.

The history of science, and more explicitly the history of the scientific revolution, can help us understand how logical rigour, experimental control, the publication of results and methods, and the very structure of scientific knowledge are not perennial facts of the history of mankind, but historical advances which can easily be lost.

A recognition of the troubled waters at the origin of modern science, an awareness that the birth of scientific learning is not quite as aseptic as the men of the Enlightenment and the positivists naively assumed does not imply either a denial of the existence of scientific knowledge, or a surrender to primitivism and the cult of magic.

Only if we are ready to renounce a portion of our childish longings, as Freud observed, can we learn to accept that some of our aspirations will turn out to be mere illusions. It seems that some students of the hermetic tradition have undergone a process not unlike that of many readers of Freud who, once they learn of the existence of the unconscious and the influence of aggressive drives at work behind the respectable veneer of civilisation and culture, come to the conclusion (unlike Freud himself) that neither reason, nor civilization, nor science exist.

Must disillusionment necessarily coincide with a desire for regression?

Magic, Metaphysics and Mysticism in the Scientific Revolution

A. RUPERT HALL

Professor Rossi's paper is of the greatest possible interest, not only because all that he writes is worthy of attention, but because in this particular case no one could accuse him of a prejudice against the trends in recent historiography which he has criticised. He is the last person of whom it might be said that his interest in intellectual history is narrow, or that his analysis is lacking in subtlety, or that he is unread in these areas whose relevance to the development of science is in question. As he mentioned himself, the subtitle of his study of Francis Bacon is "From Magic to Science" (a title used in a rather different way and long ago by Charles Singer), so that he has himself traversed some twenty years ago this very territory now so much under exploration and is certainly not without sympathy towards those aspects of human thinking which are neither inductive nor deductive, neither experimental nor mathematical. Professor Rossi has indicated his own intellectual allegiance to much of the scholarship of Miss Frances Yates and of P. M. Rattansi, while the former has expressed indebtedness to him, and so I need say no more.

Let me at once add that I am in very strong agreement with Rossi virtually all along the line; the only point on which I would differ from what he has stated—if it is indeed a point of difference at all—is that I believe that, to the special character of *historians of science* (not historians of ideas or historians of philosophy) the actual hard-line investigation of the detailed evolution of theories, experiments, measurements, that is of the solid content of scientific discourse in any period, is more important than possibly he would allow. But this issue is hardly germane to the present one.

Let me first say, however briefly, what I take the force of the paper to be. We may agree, I believe, that the traditional historiography of science, developing with increasing confidence from the beginning of the seventeenth to the beginning of the twentieth century, regarded the academic natural philosophy and medicine of earlier times as rational but mistaken and unfruitful, while it regarded the "pseudo-sciences," notably alchemy and astrology, as irrational and deceptive. The former had to be refuted by counter-argument and the marshalling of evidence for contrary hypotheses; the latter could simply be dismissed as *a priori* nonsensical. Now Professor Rossi's point is that recent philosophers, and perhaps also some recent historians, appear to place in jeopardy these twin processes of exclusion, which differ from each other in the way they work but are nevertheless alike in presuming that there is between men (even between a Ptolemy and a Copernicus, or an Aristotle and a Galileo) a common language of rational discourse. If we suppose this seeming discourse to be a farce, because the *real choice* between Ptolemy and Copernicus, or Boltzmann and Planck is made for quite other reasons, and if we hold that the magical view of nature is just as good a one to accept as any other—not better or worse than the scientific, but just different—then our interpretation of the history of science in terms of the rational discourse between men fails. It is all a tale told by an idiot. Furthermore, Rossi points out that the writings of some recent English historians of the more esoteric currents in Renaissance thought do tend directly to diminish the historical significance of rational discourse. If Copernicus wrote *De revolutionibus* because he was a Pythagorian metaphysician, what is the value of technical research into Copernicus' or Kepler's astronomical mathematics, or even into their explicit arguments for maintaining one opinion in preference to another? At this level one cannot any longer *explain* anything. Rossi raises the issue quite simply—and he is right to do so: if the history of science is concerned with rational discourse between men, then the study of alternative modes of discourse is certainly of auxiliary interest only; if on the other hand it is *not* (perhaps because of some special link between the "pseudo-sciences" and the deepest levels of the human psyche, for example), then not only has the history of science as understood for the last three hundred years been a colossal fraud, but so has science itself.

What is my reaction to all this? First, I think Rossi is quite right in supposing that this issue which I have just tried to define is one of great importance. Secondly, he is justified in claiming that some modern philosophers —totally opposing in this all their predecessors from Aristotle onwards—seek to deny the assertion that any one picture we make of the natural world at a particular moment in time is more rational than another picture. I agree with him again that the word *rational* has a definite and useful meaning; by this I mean that criteria can be established by which propositions about nature can be divided into three categories: the rational, the irrational and the doubtful. And these criteria have been accessible to all men at all times. I do not think that these criteria are culture-dependent and hence relative. Of course I do not deny that at the beginning of human intellectual history the number of doubtful propositions was large; a certain progress in geometry must be made before the search for a rational value of π becomes irrational, and, of course, although the criteria of rationality are unchanging, their application to changing sets of facts ensures that propositions irrational in one period become rational in another. In Greek antiquity it would have been irrational (that is, more precisely, counterfactual) to claim that the number of planets was greater than five, while, for us (with better observations) the opposite is true. On the other hand, the assertion: "There are really more planets than we see or shall ever see" has always been irrational.

But let us come to a case which is both historical and more difficult. I begin by quoting some words of Leibniz in an English version due to Newton.

> It may be said in a very good sense that everything is a continual miracle, that is, worthy of admiration but it seems to me that the example of a Planet which goes round and preserves its motion in its orb without any other help but that of God, being compared with a planet kept in its orb by that matter which constantly drives it towards the sun, plainly shows what difference there is between natural and rational miracles, and those that are properly so called or supernatural; or rather between a reasonable explication and a fiction invented to support an illgrounded opinion. Such is the method of those who say, after Mr. de Roberval's *Aristarchus,* that all bodies attract one another by a law of nature which God made in the beginning of things. For alleging nothing else to obtain such an effect and admitting nothing that was made by God whereby it may appear how he attains to that end, they have recourse to a miracle, that is to a supernatural thing which continues for ever, when the question is, to find out a natural cause.

The dispute between Newton and Leibniz over the explanation of gravity is well known, and I need not elaborate upon that. Notice, however, that among the issues at stake is the question: are the only explanations of phenomena that a rational science can entertain mechanical ones? Leibniz gave a firm affirmative answer; Newton's position oscillated from the doubtful to the negative. On the other hand Newton made the firm positive counterclaim that hypothetical mechanisms were not to be arbitrarily intruded into rational science. Now we could talk of antithetical paradigms of scientific theorising here, between which no smooth transition is possible; or we could speak of Newtonian canons of rationality being distinct from Leibnizian canons of rationality. But whatever language we use, let us be careful not to be impelled unthinkingly by our language into a nihilistic position, arguing that *because* these antitheses exist, there are no ultimate canons of rationality and therefore we cannot treat science as a rational discourse. The development of science itself shows this to be untrue.

For there are a number of other things to be said. In the first place, as a matter of historical fact, rational discourse between Newton and Leibniz was possible, though neither convinced the other. Clearly, despite their disputes they have something, indeed a great deal, in common. Secondly, it is not obvious that this dispute was in any necessary sense related to *science;* Newton at least was prepared to argue that although the theory of universal gravitation belonged to science, the explanation of gravity (whether by mechanical processes or by divine action) at that stage of science did not. Thirdly, the antithesis in question was not timeless but one occurring in time. Time has never resolved in some factual way the antithesis between Newton's God and Leibniz's aether; time has brought scientists to see that *in scientific terms* the proposition: God is active everywhere though we do not see him, and the proposition: the aether is active everywhere though we cannot see it, are both like the proposition: there are more planets than we can see. Time proved that the seeming antithesis between Newton and Leibniz was a pseudo-antithesis so far as natural science is concerned—whatever its other interest for us may be—because it simply had no role in the future development of celestial mechanics and cosmology.

Of course I am not maintaining that the discussion of God or the aether in the universe is not an interesting question; nor am I arguing that debate on such questions might not affect the course of subsequent research in physical science. I only say that it need not so affect it, and in this case did not. And the reason is clear: the theory of gravitation is a mathematical theory, not a metaphysical theory. Here, in a way, Newton was perfectly correct. And no very complex analysis of what happened is called for; post-Newtonian mathematical physicists simply evaded the issue of how action-at-a-distance works. But if anyone were to hold that the evasion of this particular issue (which is nevertheless a real one) indicates the irrational or arbitrary character of the historical development of science, I simply do not follow his thought.

And this brings me to Professor Rossi's distinction between historical and philosophical analysis: "Are we questioning our inadequate understanding of the genesis of modern science or modern science itself?" Obviously, the latter is strictly irrelevant to the former: I can conceive of no sort of error in the structure of modern science that would cause it to cease to be an historical phenomenon of transcendant importance. Suppose, for a moment, that science should prove in some extraordinary way to have been founded on an illusion; still, it is an illusion which has undoubtedly over several centuries increased in intellectual richness and power. It may be an old-fashioned thing to say, but I believe that you cannot study Newtonian mechanics, or Lavoisieran chemistry, or the theory of cells without discovering that here is not merely change but progress, a *change for the better* in the nature of the illusion (if science be an illusion). By *better* I mean more fertile, wider-ranging, matching more fully to experimental detail, possessing greater predictive power and so forth. Again, time's arrow points one way only: Copernicus must come before Galileo, not after; Galileo cannot succeed Newton. The historian, whatever the case with the philosopher, is not concerned simply with the way in which one view of nature comes to succeed another; he is concerned with the way in which a series of such changes is ordered in time whereby the better, the more fertile, the more subtle, the more factual comes to succeed that which is inferior in these respects. Can anyone suggest an example in the history of modern science where the *inferior* has succeeded in order of time? Let us not be bemused by the fact that the process of the succession of theories and the process of the emergence of the superior as successor both occupy time. I hold that as historians we must see the choices debated by the philosophers as both time-dependant and value-dependant; they are not things-in-themselves, everlastingly repeating, like the player's choice of a gambit in chess. Certainly if we do not take that view we run the risk of becoming incomprehensible to all but ourselves.

Lastly I would like to say a word about the growth of irrationalism. If its manifestations in the historiography of science are recent, its origins are older. One may point to such disparate manifestations as the appearance of the psychological theory of the unconscious; the writings of such (Anglo-Saxon) historians as Sir Lewis Namier and Charles Beard in which the conscious pursuit of political goals is replaced by the unrecognised drive of personal and social loyalties; Jung's interest in Paracelsus; and (if I may say so) the whole twentieth century tendency to visualize Paracelsus as one of the founding fathers of the modern scientific movement (not to mention the whole aesthetic history of our time). To record these things is not to deplore them; I merely make the point that science, for all the shocks of quantum theory, has preserved more of the rationality of our civilization than any other aspect of it. It is hardly surprising that the supreme rationality of science, so long unquestioned, should now be put in doubt. It is not, I think, particularly our business as historians to debate with those philosophers who seek to deflate the rationality of science; certainly not to argue that the present scientific world is the best of all possible worlds. But I think we may properly maintain in relation to the past that the process of the succession of theories in science was neither irrational nor random; that it was not largely directed by the workings of the unconscious, but by the conscious, reasoning mind; and that it occurred as the consequence of rational discourse between men, founded upon the knowledge of natural phenomena accessible to them.

The Mathematical Revolution of the Seventeenth Century

RENÉ TATON

During the first years immediately following the Second World War the history of science was cultivated by a few professionals, but mainly by scientists who were interested in understanding the development of their intellectual discipline, and philosophers intent on setting the evolution of scientific thought in the context of the history of philosophy. The work of Alexandre Koyré and the efforts of the school of the history of ideas to see the history of science against the background of the philosophical, religious and social ideas of a given period were relatively unknown. Hence, history of science was often considered, superficially, as the history of the warfare of reason against dogma and superstition. This naive viewpoint is illustrated in a passage from George Sarton's *The Study of the History of Science,* somewhat maliciously quoted by J. R. Ravetz in his article "Tragedy in the History of Science" (in *Changing Perspectives in the History of Science,* eds. M. Teich and R. Young, London, 1973, p. 204).

> *Definition.* Science is systematized positive knowledge, or what has been taken as such at different ages and in different places.
> *Theorem.* The acquisition and systematization of positive knowledge are the only human activities which are truly cumulative and progressive.
> *Corollary.* The history of science is the only history which can illustrate the progress of mankind. In fact, progress has no definite and unquestionable meaning in other fields than the field of science.

Ravetz has an easy time of poking fun at the dogmatic nature of these statements which, taken out of their context, describe a simplistic approach that Sarton himself never followed.

But today the danger lies elsewhere. If we still find some scientists who attempt to trace the interval history of their subject (with success, as in the case of N. Bourbaki's *Eléments d'histoire des mathématiques*), the vast majority of historians of science are well aware of the necessity of considering the history of science in the broad context of the history of civilization. Historians of science are actually pushed in this direction by an increasing number of specialists in new disciplines who stress the importance of external factors in the genesis of science. Alongside historians of philosophy, historians of ideas and philosophers of science, with whom they are familiar, historians of science now meet practitioners of such fashionable "approaches" as epistemology, the psychology of discovery, the sociology of science, the politics of science, the science of science, and so on. I do not wish to deny the value of these new fields. They frequently open new vistas and provide new light on old topics. I do wish to maintain, however, that the core of the history of science is science, itself, not its accidental paraphernalia. A host of considerations about science, its use and abuse, may be interesting and even important, but they are not essential. We should not mistake the undergrowth for the forest.

Having said this much, it is clear that I cannot accept the levelling of all stages in the history of science or the assumption that no period is more rational in its outlook than any other. If magic occasionally preceded science in certain fields and prepared its development, this does not mean that it is to be described or assessed in the same terms. It would be regrettable if the current fad for the occult, too often encouraged by irresponsible and trendy accademics, should conceal the rational and progressive character of science. If this is to be avoided, history of science must preserve its autonomy. While new disciplines should be encouraged and used, they should not be allowed to pass muster for an understanding of the nature of science, itself.

A brief survey of the evolution of the various branches of mathematics in the seventeenth century will enable us to see to what extent we can speak of a mathematical revolution on the one hand, and whether such a revolution was influenced by the hermetic tradition on the other.

During the Renaissance, mathematics was developed by various groups under varying influences. While the universities continued the medieval tradition, the Humanists were busy rediscovering the classical works of the ancient world. New methods were developed in arithmetic and algebra. Accounting flourished, and with the rediscovery of perspective a new door was open for applied mathematics. Geniuses such as Nicolas Cusanus and Leonardo da Vinci threw out suggestions that anticipated later developments.

The variety and complexity of these changes cannot be summarized in a few pages, but it is fair to say that the seventeenth century witnessed a convergence of earlier separate developments.

Let us turn first to applied mathematics. The Renaissance saw the spread of the decimal system, and at the end of the seventeenth century, Stevin introduced decimal fractions. Algebraic symbols had been invented by the German School of Rudolff and Stibel, and the extension of the possibilities of this system by Tartaglia, Cardano, Ferrari and Bombelli led to the solution of equations of the third and fourth degree. Although Alexandre Koyré had serious reservations about the role of external factors in the development of science, he recognized in his introduction to vol. II of the *Histoire Génerale des Sciences* that such factors were important in the development of elementary mathematics and influenced subsequent theory. Stevin, Viète and Descartes systematised and generalised earlier contributions, and took a new step with the introduction of more convenient notations derived from those of Nicolas Oresme in the fourteenth century and Nicolas Chuquet in the fifteenth. If there was a revolution in elementary arithmetic and algebra, it occurred in the sixteenth rather than in the seventeenth century. I believe, however, that it is more accurate to speak of an evolution that spans the period 1540–1640.

Besides the innovation represented by the introduction of algebra, two other major developments must be mentioned.

The first is the invention of logarithms by Napier and Bürgi at the beginning of the seventeenth century. These were to provide the indispensable tools for the new calculations required in astronomy and navigation. The second is the application of algebra to geometry by Descartes and Fermat. The discovery of analytical geometry is only in a very general sense in the line of ideas that were discussed between Apollonius and Oresme. It is mainly the outcome of a quest for a new "calculus" that could provide a basis for the "*mathesis universalis*" that was to encompass the whole of nature. The notion of function was introduced by means of the graph and proved of revolutionary importance for subsequent developments.

A. P. Youschkevitch, at the Prague Symposium of September 1967, recalled that F. Engels had stressed the fundamental importance of Descartes' contribution. The notions of function and variable were not unheralded, but it is in analytical geometry that they were successfully applied. By the end of the seventeenth century, Leibniz and Jean Bernouilli will already be introducing relatively clear and rigorous definitions of the new technique.

The determination of surfaces, volumes and centres of gravity by what the seventeenth century called "the method of exhaustion" had been introduced by Archimedes in the third century B.C. After a long period of neglect, this method was revised in the second half of the seventeenth century after the publication in 1544 of the *editio princeps* of Archimedes' works. The contribution of Luca Valerio, and certain developments made by Galileo and Kepler are the forerunners of Cavalieri's geometry of indivisibles, a new realm explored by Grégoire de Saint Vincent, Roberval, Torricelli, Pascal, Wallis, Gregory, Huygens, Stevin and others. The culmination is the creation of the algorithms of infinitesimal calculus by Newton and Leibniz in the last quarter of the seventeenth century. This work combines the principles of integral calculus (underlying the problems of indivisibles discussed by Cavalieri) with the principles of differential calculus used to solve the problem of tangents and the determination of speeds that Galileo and Descartes thrust to the fore.

The power of the new method was displayed at the end of the seventeenth and during the eighteenth century in the solution of difficult problems. Thanks to differential equations, the hope expressed by Galileo half a century earlier was becoming real by the end of the seventeenth century: Nature proved to be written in mathematical language. With the notion of function and analytical geometry, the new calculus enabled mathematicians to pursue the theoretical study of the most fundamental phenomena of physics and to envisage the extension of mathematics to all the branches of physical science. We can speak of two turning points around the years 1630 and 1680. The essential aspects are the introduction of the concept of function, the development of differential and integral calculus, and the discovery of single and powerful algorithms which were to be the basis of Leibniz's dream of a universal symbolic language. These developments were influenced by practical concerns and problems in physics, but the essential drive seems to have been internal to mathematics, itself.

Linked to elementary algebra by their common source, the *Arithmetic* of Diophantes, the theory of numbers is a seventeenth century innovation due to Fermat, although it was only developed during the next century by Euler and Lagrange. It is noteworthy that Fermat refers only to Diophantes and Archimedes and never to the neoplatonic arithmological speculation that was common in the sixteenth century and influenced Kepler.

The calculus of probabilities dates from 1654 when Pascal and Fermat exchanged letters on certain aspects of gambling. E. Courmet has shown in a recent doctoral dissertation, *Mersenne, Frénicle et l'élaboration de l'analyse combinatoire dans la prémière moitié du XVIIe siècle,* that the origins of this calculus are complex and include games of chance, juridical and theological questions, and cabilitistic and hermetic elements. The correspondence of Mersenne bears witness to the network of problems and questions that influenced the development of combinatory analysis. But it is probably earlier in this field that we can speak of a direct influence of the hermetic tradition. Pascal called his famous hexagon "mystical", and some of the considerations he makes prove that he was familiar with the terminology and ideas of the occult philosophy.

In geometry the main effort is directed towards the rediscovery of the Greek sources, especially the works of Euclid, Apollonius and Pappus. The admiration for the richness, elegance and vigour of these texts was so great that the seventeenth century is mainly a period of imitation and assimilation. Nevertheless, the theory of perspective and applied geometry was expanded to meet the needs of the rising class of engineers and artists. From the time of Leon Battista Alberti and Leonardo da Vinci in the fifteenth century, to Dürer and Stevin in the sixteenth, the theory of perspective made rapid progress, but it only became a formalized science in 1639 when R. Descartes created prospective geometry by audaciously combining classical concepts with the practical methods of draughtsmen.[1] This geometry was taken into consideration by Pascal and Philippe de la Hire, but it was soon forgotten until it was rediscovered by

Jean-Victor Poncelet in 1820. The reason for this may be quite simply that the best minds were too busy with analytical geometry and calculus to have time for much else.

Drawing together the threads of this rapid outline, we can say that there were two turning points in the development of mathematics. The first occurred in the 1630's with the systematization of elementary algebra, the creation of analytical geometry, prospective geometry and the calculus of indivisibles. The second happened in the 1680's with the algorithmisation of infinitesimal calculus. But the movement must be seen as continuous rather than discrete: there was no one instantaneous revolution in mathematics but a steady transformation of the field during the last two thirds of the seventeenth century.

D. T. Whiteside believes that the causes of this transformation are internal to mathematics itself, for "by the 17th century mathematical structure had become too systematised and too remote from any possible physical origins to allow any further incursion of concepts from without". A. P. Youschkevitch, however, feels that this viewpoint is unsatisfactory:

> *Au XIIe siècle, le processus de la connaissance du monde physique avait lieu tant dans la formation des structures mathématiques, correspondant à l'ensemble des sciences physiques, nouvellement créée, que dans celle des structures de ces sciences physiques, sur des bases empiriques et mathématiques, avant tout de la mécanique rationnelle des mouvements terrestres et célestes. Au cours de l'interaction entre la mathématique et les autres sciences du XIIe siècle . . . les représentations physiques étaient traitées en concepts mathématiques ou bien stimulaient leur formation, tandis que la mathématique contribuait à expliquer et à formuler les principales notions physiques.*[3]

These relations between the development of mathematics and the development of science are easier to understand when we recollect that seventeenth century geometers were often, as well, physicists, astronomers, physicians or philosophers. They were intent not on developing purely abstract systems but on coping with the new problems at hand. Hermeticism may have influenced their research, but this is a secondary factor compared to the importance of the general philosophical climate and the pressure exerted by the new technological requirements.

NOTES AND REFERENCES

The Chemical Debates of the Seventeenth Century: The Reaction to Robert Fludd and Jean Baptiste van Helmont

ALLEN G. DEBUS

The present paper is based upon several chapters of my forthcoming book, *The Chemical Philosophy*. The following references are based upon the more detailed notes to be found there.

[1] Representative seventeenth century definitions will be found in Nicholas Lemery, *A Course of Chymistry . . .*, trans. Walter Harris, M.D. (2nd English edition from the 5th French edition, London: Walter Kettilby, 1686), p. 2; Nicholas le Febure (Le Févre), *A Compleat Body of Chymistry . . .*, trans. P.D.C., Esq. (Corrected edition, London: O. Pulleyn for John Wright, 1670), Part I, pp. 6–12. Additional pertinent material is to be found in the present author's "Alchemy" in the *Dictionary of the History of Ideas,* Executive editor, Philip P. Wiener (New York: Charles Scribner's Sons, 1973), *1,* pp. 27–34 and "Renaissance Chemistry and the Work of Robert Fludd," *Ambix, 14* (1967) 42–59.

[2] The influence of *Ecclesiasticus* 38 is evident in Paracelsus' contemptuous address to the physicians of his day [from the *Paragranum,* see Paracelsus, *Sämtliche Werke,* ed. Karl Sudhoff and Wilhelm Matthiessen (15 vols., Munich and Berlin, 1922–1933), *8,* pp. 63–65] and in Jean Baptiste van Helmont's lengthy discussion of the text offered in an autobiographical context ["De Lithiasi" ("Philiatro Lectori") in *Opuscula Medica Inaudita* (Amsterdam: Ludovicus Elsevir, 1648; reprinted Brussels: Culture et Civilisation, 1966), p. 4; and in English translation in the *Oriatrike or Physicke Refined. The Common Errors therein Refuted, And the Whole Art Reformed & Rectified. Being a New Rise and Progress of Phylosophy and Medicine, for the Destruction of Diseases and Prolongation of Life,* trans. J(ohn) C(handler) (London: Lodowick Loyd, 1662), sig. Nnnnn 2v].

[3] Traditional alchemical claims regarding the scope of the science remained influential in the seventeenth century. Van Helmont was to quote the 14th century *pseudo*-Lullius at length on this point: ". . .our *Philosophers* or followers, have directed themselves to enter through any kind of Science, into all experience, by Art, according to the course of Nature in its Univocal or single Principles. For *Alchymie* alone, is the Glass of true understanding; and shews how to touch, and see the truths of those things in the clear Light." Van Helmont, "De Lithiasi" (Chap. 3, Sect. 1), *Opuscula Medica Inaudita* (Frankfurt: Joh. Just. Erythropili, 1682), p. 12; *Oriatrike or Physick Refined,* pp. 839–840.

[4] This is discussed at some length by the present author in "Mathematics and Nature in the Chemical Texts of the Renaissance," *Ambix, 15* (1968), 1–28.

[5] Again this is a persistent theme in the literature—perhaps most succinctly stated by the English Paracelsian, Thomas Tymme: "The almighty Creatour of the Heauens

and Earth (Christian Reader), hath set before our eyes two most principall Bookes: the one of Nature, the other of his written Word. . ." *A Dialogue Philosophicall* (London, 1612), sig. A3 (from the dedication to Sir Edward Coke, Lord Chief Justice of the Court of Common Pleas).

[6] Note the anonymous author of the *Philiatros* exhorting his readers to "put then on Gloues and Cuffes, for you must to the fire, and happily to the fiery Furnace." (London, 1615), fol. 14. And, frequently repeated was Peter Severinus' statement that true physicians should discard their books, collect samples of all things in nature and subject them to the fire in their laboratories. *Idea Medicinae Philosophicae* (1571; 3rd ed., Hagae Comitis: Adrian Clacq, 1660), p. 39.

[7] Paracelsus, *Labyrinthus medicorum* in *Opera Omnia Medico-Chemico-Chirurgica,* trans. F. Bitiskius (3 vols., Geneva: Ioan. Antonij, & Samuelis De Tournes, 1658), *1,* pp. 264–288 (275).

[8] Paracelsus' doctrine of the microcosm is discussed in the *Philosophia Sagax* [*Opera Omnia* (Bitiskius ed., 1658), *2,* p. 601] and is described by Walter Pagel, *Paracelsus. An Introduction to Philosophical Medicine in the Era of the Renaissance* (Basel/New York: S. Karger, 1958), pp. 65–68.

[9] On this see Allen G. Debus, "Edward Jorden and the Fermentation of the Metals: An Iatrochemical Study of Terrestrial Phenomena" in *Toward a History of Geology,* ed. Cecil J. Schneer (Cambridge, Mass.: M.I.T. Press, 1969), pp. 100–121.

[10] This is a prominent theme in Paracelsus, *Philosophia ad Atheniensis* (1564) and in Gerhard Dorn, *Liber de Naturae luce Physica, ex Genesi desumpta* (1583). In England Thomas Tymme insisted that "Halchymie should have concurrence and antiquitie with Theologie," since Moses "tels us that the Spirit of God moved upon the water: which was an indigested Chaos or masse created before by God, with confused Earth in mixture: yet, by his Halchymicall Extraction, Seperation, Sublimation, and Coniunction, so ordered and conioyned againe, as they are manifestly seene a part and sundered: in Earth, Fyer included, (which is a third Element) and Ayre, [and] (a fourth) in Water. . ." From the dedication to Sir Charles Blunt in Joseph Duchesne (Quercetanus), *The Practise of Chymicall, and Hermeticall Physicke, for the preseruation of health,* trans. Thomas Tymme (London: Thomas Creede, 1605), sig. A3 *r.*

[11] The "chemical" role of the Paracelsian *archeus* is developed in Paracelsus, *Volumen Medicinae Paramirum,* trans. and with a preface by Kurt F. Leidecker, *Supp. Bull. Hist. Med.,* No. 11 (Baltimore: Johns Hopkins Press, 1949), p. 29. For a later period see Audrey B. Davis, *Circulation Physiology and Medical Chemistry in England 1650-1680* (Lawrence, Kansas: Coronado Press, 1973).

[12] See Allen G. Debus, *The English Paracelsians* (London: Oldbourne Press, 1965), pp. 26–29 and 45.

[13] Note the ongoing debate over this issue described in Allen G. Debus, "The Paracelsians and the Chemists: the Chemical Dilemma in Renaissance Medicine," *Clio Medica,* 7 (1972), 185–199.

[14] Thomas Erastus, *Disputationes de Medicina Nova Paracelsi* (4 parts, Basel, 1572–1574). The debate in Paris in 1579 is discussed by Hugh Trevor-Roper, "The Sieur de la Rivière, Paracelsian physician of Henry IV," *Science, Medicine and Society in*

the Renaissance, ed. Allen G. Debus (2 vols., New York: Science History Publications, 1972), *2,* pp. 227–250. See also Dietlinde Goltz, "Die Paracelsisten und die Sprache," *Sudhoffs Archiv, 56,* 337–352. The early Paracelsian debates are discussed by the present author in a book-length manuscript now nearing completion, *The Chemical Philosophy: Paracelsian Science and Medicine in the Sixteenth and Seventeenth Centuries* (Chapter 3: "The Paracelsian Debates").

[15] The most recent account of the Rosicrucians will be found in Frances A. Yates, *The Rosicrucian Enlightenment* (London/Boston: Routledge & Kegan Paul, 1972). She discusses the voluminous literature on this topic, but her conclusions are not always proven. See also Allen G. Debus, *Science and Education in the Seventeenth Century. The Webster-Ward Debate* (London: Macdonald/New York: American Elsevier, 1970), pp. 15–32 *passim.*

[16] In recent years Robert Fludd has been the subject of numerous studies by Frances A. Yates, C. H. Josten, Serge Hutin, the present author and others. This literature is summarized in Allen G. Debus, "Robert Fludd," *Dictionary of Scientific Biography,* Editor-in-Chief, Charles C. Gillispie (New York: Charles Scribner's Sons, 1972), *5,* pp. 47–49. The standard biography remains that of J. B. Craven [*Doctor Robert Fludd (Robertus de Fluctibus). The English Rosicrucian. Life and Writings* (Kirkwall, 1902; reprinted New York, n.d.)]. Although badly outdated, this still contains useful material.

[17] Robert Fludd, *Apologia Compendiaria Fraternitatem de Rosea Cruce Suspicionis et Infamiae Maculis Aspersam, Veritatis quasi Fluctibus abluens et abstergens* (Leiden: Godfrid Basson, 1616). The second—greatly enlarged—edition, *Tractatus Apologeticus Integritatem Societatis De Rosea Cruce defendens* (Leiden: Godfrid Basson, 1617) was used as the basis of the following account. Here see especially pp. 91–124. The present account of the work and debates of Fludd is based upon the fourth chapter of my forthcoming *The Chemical Philosophy.*

[18] Fludd, *Tractatus Apologeticus,* pp. 187–192.

[19] While we may call Fludd an "atomist" at this date [see also "The sixt experiment wch maketh it probable y^t all things wer made of Atoms as some Philosophers haue gessed" (fol. 79r of Fludd's *A Philosophicall Key,* Trinity College, Cambridge, Western MS. 1150 (0.2.46). The present author is preparing an edition of this manuscript for publication] it is important to note that "corpuscularian" explanations did not play a major part of his system. Important recent discussions of early seventeenth century atomism are to be found in J. E. McGuire and P. M. Rattansi, "Newton and the Pipes of Pan," *Notes and Records of the Royal Society, 21* (1966), 108–141 and in the series of papers by T. Gregory, "Studi sull'atomismo del seicento," *Giornale Critico della Filosofia Italiana, 43* (1964), 38–65; *45* (1966), 44–63; *46* (1967), 528–541. Here the section on the early seventeenth century iatrochemist, Daniel Sennert (1572–1637) is of special interest [*45* (1966), 51–63].

[20] Fludd, *Tractatus Apologeticus,* pp. 190–192.

[21] Robert Fludd, *Utriusque Cosmi Maioris scilicet et Minoris Metaphysica, Physica atque Technica* (Oppenheim: T. de Bry, 1617).

[22] Among the many accounts of the Fludd-Kepler debate the most detailed is still that of Wolfgang Pauli, "The Influence of Archetypal Ideas on the Scientific Theories of

Kepler" in C. G. Jung and W. Pauli, *The Interpretation of Nature and the Psyche,* trans. Priscilla Silz (New York: Pantheon Books, Bollingen Series 51, 1955), pp. 147–240.

[23] Johannes Kepler, *Adversus Demonstrationem Analyticam . . . de Fluctibus* in *Gesammelte Werke* (18 vols., München: C. H. Beck, 1937–1949), *6* (1940), p. 431.

[24] Kepler, *Harmonices Mundi, Appendix habet comparationem huius Operis cum Harmonices Cl. Ptolemaei libro III cumque Roberti de Fluctibus, dicti Flud. Medici Oxoniensis speculationibus Harmonicis, operi de Macrocosmo & Microcosmo insertis* in *Ibid.,* p. 374.

[25] Kepler, *Adversus Demonstrationem Analyticam . . . de Fluctibus . . .* in *Ibid.,* p. 428.

[26] As quoted by Pauli, *Op. cit.,* p. 196 from the *Veritatis Proscenium,* p. 12.

[27] See Gabriel Naudé, *Instruction à la France sur la Verité de l'Histoire des Frères de la Roze-Croix* (Paris: François Iulliot, 1623), sig. c̃v and p. 27.

[28] This episode is discussed in a lengthy note by the editor in the *Correspondance du P. Marin Mersenne. Religieux Minime, Publiée par Mme. Paul Tannery, Editée et annotée par Cornelis de Waard avec la collaboration de René Pintard,* I (1617–1627) (Paris: P.U.F., 1945), pp. 167–168. The theses are described by Mersenne in *La Vérité des Sciences. Contre les Septiques ou Pyrrhoniens* [Paris: Toussainct Du Bray, 1625; reprint Stuttgart-Bad Canstatt: Friedrich Fromman Verlag (Günther Holzboog), 1969], pp. 79–80.

[29] Mersenne, *La Vérité des Sciences,* p. 41.

[30] *Ibid.,* pp. 81–83.

[31] *Ibid.,* p. 56.

[32] *Ibid.,* pp. 105–107.

[33] *Ibid.,* pp. 116–119.

[34] Mersenne, *Correspondance, 1,* p. 62.

[35] Robert Fludd, *Sophiae cum Moria Certamen, Inquo, Lapis Lydius a Falso Structore, Fr. Marino Mersenno, Monacho, Reprobatus, celeberrima Voluminis sui Babylonici (in Genesin) figmenta accurate examinat* [s.l. (Frankfurt), 1629]; Joachim Frizius (Fludd?), *Summum Bonum, Quod est Verum [(Magiae, Cabalae, Alchymiae: Verae); Fratrum Roseae Crucis verorum] subjectum. . .* [(Frankfurt?), 1629].

[36] *Ibid.,* p. 31.

[37] Pierre Gassendi, *Examen Philosophiae Roberti Fluddi* in *Opera Omnia* (6 vols., Florence, 1727), *3,* pp. 224–267. On the Fludd-Gassendi debate see Luca Cafiero, "Robert Fludd e la Polemica con Gassendi", *Rivista Critica di Storia della Filosofia, 19* (1964), 367–410; *20* (1965), 3–15; T. Gregory, *Scettisismo ed empirismo: Studio su Gassendi* (Bari: Laterza, 1961), pp. 50–62; Robert Lenoble, *Mersenne ou la naissance du mécanisme* (Paris: Vrin, 2nd ed., 1971), *passim;* Frances A. Yates, *Giordano Bruno and the Hermetic Tradition* (Chicago: University of Chicago Press, 1964), pp. 432–455.

[38] See Allen G. Debus, "Robert Fludd and the Circulation of the Blood", *Journal of the History of Medicine and Allied Sciences, 16* (1961), 374–393 and "Harvey and Fludd: The Irrational Factor in the Rational Science of the Seventeenth Century", *Journal of the History of Biology, 3* (1970), 81–105.

[39] Pierre Gassendi, *Epistolica exercitatio in qua principia philosophiae Roberti Fluddi, medici, reteguntur, et ad recentes illius libros adversus R.P.F. Marinum Mersennum . . . respondetur* (Paris, 1630), pp. 133–136.

[40] Robert Fludd, *Clavis philosophiae et alchymiae Fluddanae, sive Roberti Fluddi Armigeri, ut medicinae doctoris, ad epistolicam Petri Gassendi theologi exercitatem responsum* (Frankfurt: Wilhelm Fitzer, 1633), pp. 33ff.

[41] P. Jean Durelle, *Effigies contracta Roberti Flud Medici Angli, cum Naevis, Appendice et Relectione. In lucem producente Eusebio a S. Justo Theologo Segusiano. Sanctitas Dei defensa. Agnito vultus eorum respondebit Isa. 3. Ad Clariss, virum Iacobum Dazam* (Paris: Apud Guillelmum Baudry, 1636).

[42] The present account is based upon the fifth chapter of my manuscript, *The Chemical Philosophy.* The Helmontian literature has been discussed frequently by Walter Pagel. His most recent—and most convenient—description will be found in his article on van Helmont in the *Dictionary of Scientific Biography, 6* (1972), pp. 253–259 (257–259). This is to be supplemented with the bibliography of secondary sources prepared by J. R. Partington, *A History of Chemistry* (London: Macmillan, 1961), *2* pp. 209–210 (note 4).

[43] Van Helmont (Brussels) to Mersenne (Paris), 19 December 1630, in the *Correspondance, 2,* pp. 582–599 (584).

[44] J. B. van Helmont, *Ortus Medicinae. Id est, Initia Physicae Inaudita. Progressus medicinae novus, in Morbrrum Ultionem, ad Vitam Longam* (Amsterdam: Ludovicus Elzevir, 1648; reprinted Brussels: Culture et Civilisation, 1966); *"Confessio Authoris"* (Chap. 2, sects. 1–5), p. 16, *Oriatrike* (1662), pp. 11–12.

[45] This subject is discussed in some detail in Allen G. Debus, "Robert Fludd and the Use of Gilbert's *De Magnete* in the Weapon-Salve Controversy", *Journal of the History of Medicine and Allied Sciences, 19* (1964), 389–417.

[46] *"De Magnetica Vulnerum Curatione"* (sect. 1), *Ortus medicinae* (1648), pp. 748–750; *Oriatrike* (1662), pp. 759–761.

[47] *"De Magnetica Vulnerum Curatione"* (sect. 130), *Ortus medicinae* (1648), p. 772; *Oriatrike* (1662), p. 785.

[48] *"De Magnetica Vulnerum Curatione"* (sect. 126), *Ortus medicinae* (1648), p. 771; (sect. 125 in English ed.) *Oriatrike* (1662), p. 784.

[49] *"De Magnetica Vulnerum Curatione"* (sect. 174), *Ortus medicinae* (1648), p. 780; (sect. 173 in English ed.) *Oriatrike* (1662), p. 793.

[50] The complex history of the prosecution is succinctly discussed by Nève de Mévergnies, *Jean-Baptiste van Helmont. Philosophe par le feu* (Paris: Librairie E. Droz, 1935), pp. 123–143. It should be noted that this account is deeply indebted to the earlier work of C. Broeckx and A. J. J. Vandevelde.

[51] The relevant texts are J. B. van Helmont, *Paradoxa de aquis Spadanis fontibus* and *Supplementum de Spadanis fontibus* (Leodii: L. Streel, 1624) and Henricus van Heer, *Spadacrene, hoc est Fons Spadanus: ejus singularia, bibendi modus, medicamina bibentibus necessaria* (Leodii: Arn. de Corswaremia, 1614). The work by van Heer went through many editions throughout the seventeenth century.

[52] Van Helmont (Brussels) to Mersenne (Paris), 14 February 1631, *Correspondance, 3,* pp. 95–109 (108).

[53] "... *alter pyrotechnice philosophando, has plus quam cimmerias tenebras, quibus utriusque intellectus altissime obsessus est, et pro nutu daemonis circumagitur, toti mundo effundant*. ..." As quoted by Nève de Mévergnies *op. cit.,* p. 133.

[54] *"Promissa Authoris", Ortus medicinae* (1648), p. 6; *Oriatrike* (1662), p. 1. See also the *"Logica inutilis"* (sect. 3), *Ortus medicinae* (1648), p. 41; *Oriatrike* (1662), p. 37. Here he consciously refers to a *"nova Philosophia".*

[55] The Helmontian views on motion are discussed at length in Allen G. Debus, "Motion in the Chemical Texts of the Renaissance", *Isis, 64* (1973), 4–17.

[56] *"Blas Humanum"* (sect. 3), *Ortus medicinae* (1648), p. 180; *Oriatrike* (1662), p. 176.

[57] *"Causae, et Initia Naturalium"* (sect. 41), *Ortus medicinae* (1648), p. 39; *Oriatrike* (1662), p. 34.

[58] *"Pharmacopolium ac Dispensatorium modernum"* (sect. 32), *Ortus medicinae* (1648), p. 463; *Oriatrike* (1662), p. 462.

[59] Van Helmont (Brussels) to Mersenne (Paris), 30 January 1631, *Correspondance 3,* pp. 56–57. Both the willow tree experiment and the suggestion that a study of the comparative weights of metals might be made are to be found also in the fourth book of the *Idiota* of Nicholas of Cusa.

[60] *"Scholarum Humoristorum passiva Deceptio atque Ignorantia"* (Chap. 4, sect. 31), *Opuscula Medica Inaudita* (1648), p. 108; *Oriatrike* (1662), p. 1056. Once again, in the *Idiota,* Cusanus suggested that a physician might give a more valid judgment of urine "by the weight and colour both together then by the deceitfull colour alone". Cusanus, *The Idiot in Four Books* ... (London: William Leake, 1650), p. 174.

[61] *"Vacuum naturae"* (Sects. 7–11), *Ortus medicinae* (1648), pp. 84–85; *Oriatrike* (1662), pp. 82–83.

[62] *"Imago fermenti etc."* (sect. 33), *Ortus medicinae* (1648), p. 117; *Oriatrike* (1662), p. 117.

[63] *"Aura Vitalis", Ortus medicinae* (1648), p. 726; *Oriatrike* (1662), p. 733; *"Spiritus Vitae"* (sect. 15–16), *Ortus medicinae* (1648), pp. 198–199; *Oriatrike* (1662), p. 195. The influence of this work on Robert Boyle has been touched on by A. R. Hall in his "Medicine and the Royal Society", *Medicine in Seventeenth Century England,* ed. Allen G. Debus (Berkeley: University of California Press, 1974), pp. 421–452.

[64] *"De Febribus"* (chap. 4), *Opuscula Medica Inaudita* (1648), pp. 17–25; *Oriatrike* (1662), pp. 949–957. The subject has been covered intensively by P. H. Niebyl in "Galen, van Helmont and Blood Letting", *Science, Medicine and Society in the Renaissance,* ed. Allen G. Debus, *2,* pp. 13–23.

[65] *"Sextuplex Digestio. . ."* (sect. 56), *Ortus medicinae* (1648), p. 220; *Oriatrike* (1662), p. 217; *"De Lithiasi"* (Chap. 5, sect. 17), *Opuscula Medica Inaudita* (1648), p. 45; *Oriatrike* (1662), p. 863; *"Potestas medicaminum"* (sects. 36–37), *Ortus medicinae* (1648), p. 479; *Oriatrike* (1662), pp. 477–478.

[66] Allen G. Debus, "Palissy, Plat and English Agricultural Chemistry in the 16th and 17th Centuries", *Archives Internationales d'Histoire des Sciences, 21* (1968), 67–88. In this regard Johann Rudolph Glauber's *Des Teutschlandts Wolfahrt* (1656–1661) is of considerable importance and it is likely that a study of J. J. Becher's economic program in the context of the chemical philosophy would prove a fruitful field of research.

[67] J. B. van Helmont, *Ternary of Paradoxes, The Magnetick Cure of Wounds. Nativity of Tartar in Wine, Image of God in Man,* Translated, Illustrated and Ampliated by Walter Charleton (London: James Flesher for William Lee, 1650). See also Nina Rattner Gelbart, "The Intellectual Development of Walter Charleton", *Ambix, 18* (1971), 149–168.

[68] Allen G. Debus, "The Paracelsian Aerial Niter," *Isis, 55* (1964), 43–61.

[69] This is most evident in the *Sceptical Chymist* (1661 but composed in the early 1650s) and in the first part of *Some Considerations Touching the Usefulness of Experimental Natural Philosophy* (1663 but composed c. 1648–1649). Boyle's investigation of neutralization in acid-base reactions (1675) forms part of the Helmontian tradition while he was still to quote van Helmont as a deciding authority in his *Natural History of the Human Blood* (1685). On Boyle's indebtedness to the older authors in regard to analysis see Allen G. Debus, "Fire Analysis and the Elements in the Sixteenth and the Seventeenth Centuries," *Annals of Science, 23* (1967), 127–147.

[70] Thomas Willis' *De Fermentatione* (1659) is strongly Helmontian in tone. Although an eclectic, the Helmontian sources of Willis are quite evident and have been properly identified by Hansruedi Isler in his *Thomas Willis 1621–1675. Doctor and Scientist* (New York/London: Hafner, 1968). On this see also Audrey Davis, *op. cit.* An excellent recent account of Sylvius will be found in Lester S. King, *Road to Medical Enlightenment 1650–1695* (New York: American Elsevier/London: Macdonald, 1970), pp. 93–112. See also Jose Maria López Piñero, "*La Iatroquimica de la Segunda Mitad del Siglo* XVII" in *Historia de la Medicina,* ed. Pedro Laín Entralgo (Barcelona: Salvat Editores, 1973), *4,* pp. 279–295.

[71] "Natura Contrarium Nescia" (sect. 11 *seq.*), *Ortus medicinae* (1648), p. 167; *Oriatrike* (1662), pp. 163–164. See above, note 55.

[72] "Ignota Actio Regiminis" (sect. 3), *Ortus medicinae* (1648), p. 329; *Oriatrike* (1662), p. 325. The Latin for this important passage reads "*itaque stabilivere omne patiens, vicissim reagere necessario, atque eatenus similiter omne Agens repati, nec etiam aliunde debilitari*". This is a far cry from Newton's succinct statement of the third law of motion, "*Actioni contrarium semper & aequalem esse reactionem: sive corporum duorum actiones in se mutuo semper esse aequales & in partes contrarias dirigi*". Isaac Newton, *Philosophiae Naturalis Principia Mathematica* (London: Jussu Societatis Regiae ac Typis Josephi Streater, 1687), p. 13. A paper on the relationship of Helmont to Newton and the third law is forthcoming from the present author.

[73] Van Helmont, "*Ignota Actio Regiminis*" (sect. 4), *Ortus medicinae* (1648), p. 329; *Oriatrike* (1662), p. 325.

[74] "*Ignota Actio Regiminis*" (sect. 6), *Ortus medicinae* (1648), p. 332; *Oriatrike* (1662), p. 327.

[75] *"Ignota Actio Regiminis"* (sect. 7), *Ortus medicinae* (1648), p. 332; *Oriatrike* (1662), p. 327.

[76] The Newton manscript (King's College, Cambridge, Keynes, Newton MS 16) begins "pag. 21. *Causae et initia naturalium"* and closes with notes on the *"Formarum ortus"*. The page references are to the 1667 edition of the *Opera omnia.* Additional evidence of Newton's interest in van Helmont may be found in *A Catalogue of the Portsmouth Collection of Books and Papers Written by or belonging to Sir Isaac Newton...* (Cambridge: Cambridge University Press, 1888). Here, under Sec. II: Chemistry, see p. 13, No. 8, "Extracts apparently from Van Helmont", and p. 16, No. 4, *"De Peste.* Van Helmont".

[77] Mary Hesse, "Reasons and Evaluation in the History of Science" in *Changing Perspectives in the History of Science. Essays in Honour of Joseph Needham,* eds. Mikulás Teich and Robert Young (London: Heinemann, 1973), pp. 127–147 (143).

The author wishes to express his gratitude to the National Science Foundation for research support during the period of preparation of the present paper (research grant number GS-37063).

New Light on Galileo's Lunar Observations

GUGLIELMO RIGHINI

[1] Galileo, *Dialogo sopra i due massimi sistemi* [1632] in *Opere di Galileo Galilei,* ed. A. Favaro, 20 vols., Florence: Barbèra 1890–1909, VII, 90. English translation by Stillman Drake, *Dialogue Concerning the Two World Systems,* Berkeley and Los Angeles: University of California Press, 1967, pp. 65–66.

[2] *Opere di Galileo,* VII, 91; Drake trans. in *Dialogue,* pp. 66–67.

[3] *Opere di Galileo,* XVII, 211, 214–215.

[4] *Opere di Galileo,* XVII, 275.

[5] *Opere di Galileo,* III, 73. English translation by Stillman Drake in *Discoveries and Opinions of Galileo,* Garden City: Doubleday Anchor Books, 1957, p. 42.

[6] *Opere di Galileo,* III, 75; *Discoveries and Opinions,* p. 45.

[7] *Opere di Galileo,* VII, 93; Drake trans. in *Dialogue,* p. 69.

[8] *Opere di Galileo,* VII, 93; Drake trans. in *Dialogue,* p. 98.

[9] J. Classen, "The First Maps of the Moon", *Sky and Telescope 37* (1969), 82.

[10] Z. Kopal, *The Moon,* Dordrecht: D. Reidel, 1969, p. 225.

[11] *Opere di Galileo,* X, 250, 253.

[12] *Opere di Galileo,* X, 277.

[13] I have used P. Ahnert's *Astronomische Chronologische Tafeln,* 2nd edition, Leipzig: Johann Ambrosius Borth, 1961.

Dissertatio cum Professore Righini et Sidereo Nuncio

OWEN GINGERICH

[1] See especially N. R. Hanson, *Patterns of Discovery* (Cambridge, 1965).

[2] Donald deB. Beaver, "Bernard Walther: Innovator in Astronomical Observation," *Journal for the History of Astronomy 1,* 39–43, 1970.

[3] Codex Barbarensis 39, folio 55, Vatican Library; quoted by Joseph Hilgers, *Der Index des verbotener Bücher* (Freiburg, 1904), pp. 541–542.

[4] Copernicus observations are conveniently listed by Noel Swerdlow in an essay review, "The Holograph of *De revolutionibus* and the Chronology of its Composition," *Journal for the History of Astronomy,* 5, 186–198, 1974.

[5] Owen Gingerich, "Remarks on Copernicus' Observations," in Robert Westman, ed., *The Copernican Achievement* (Berkeley-Los Angeles, 1975).

[6] Waldemar Voisé, "The Great Renaissance Scholar," pp. 84–94, esp. p. 94 in B. Bieńkowska, ed., *The Scientific World of Copernicus* (Dordrecht-Boston, 1973).

[7] Victor E. Thoren, "Tycho and Kepler on the Lunar Theory," *Publications of the Astronomical Society of the Pacific, 79,* 482–489, 1967.

[8] Translated from Johannes Kepler, *Astronomia nova,* chapter 11.

[9] William R. Shea, "Galileo, Scheiner, and the Interpretation of Sunspots," *Isis 61,* 498–519, 1970.

[10] Dora Shapley, "Pre-Huygenian Observations of Saturn's Ring," *Isis 40,* 12–17, 1949; Albert van Helden, "Saturn and his Anses," *Journal for the History of Astronomy, 5,* 105–121, 1974.

[11] A. Favaro, ed., *Le Opere di Galileo Galilei,* vol. 3, part 2 (Florence, 1907), pp. 427 ff.

[12] *Ibid.,* vol. 3 part 1 (Florence, 1892) pp. 35 ff.

[13] *Ibid.,* vol. 3, part 1, p. 48.

[14] Jean Meeus, "Galileo's First Records of Jupiter's Satellites," *Sky and Telescope 27,* 105–106, 1964.

[15] Translated in Stillman Drake, *Discoveries and Opinions of Galileo* (Garden City, New York, 1957), pp. 34 and 36.

Marcello Malpighi and the Founding of Anatomical Microscopy

LUIGI BELLONI

[1]H. B. ADELMANN, *Marcello Malpighi and the evolution of embryology*, Ithaca, N.Y., Cornell University Press, 1966.

[2]Opere scelte di *Marcello Malpighi a cura di Luigi Belloni,* Torino, U.T.E.T., 1967. This volume is illustrated with photographs of preparations made according to the techniques used by Malpighi (see my article, "The Repetition of Experiments and Observations: Its Value in Studying the History of Medicine [and Science,]" in *Journal of the History of Medicine,* 25 [1970], 158–168).

Works published after 1967 pertaining to Malpighi and his influence, to the history of anatomical microscopy and to the importance of the Galilean school:

"*Il primo ventennio della microscopia (Galilei 1610–Harvey 1628). Dalla microscopia alla anatomia microscopica dell'insetto,*" in *Clio Medica,* 4 (1969), 179–190.

"*Bionica del palombaro e del sommergibile dal Galilei al Borelli,*" in *Simposi Clinici,* 7 (1970), XVII–XXIV.

"Italian Medical Education after 1600," in *The History of Medical Education* (C. D. O'Malley, Ed.), University of California Press, Berkeley-Los Angeles-London, 1970, 105–119.

"*Übereinstimmungen zwischen Stensen und Malpighi auf dem Gebiet der Sekretion und der Inkrustation,*" in *Dissertations on Steno as Geologist* (Ed. G. Scherz), Odense University Press, 1971, 140–148.

"*Auf dem Wege zur Gewebepathologie: 1769 erkennt Cotugno die schichteigenen Veränderungen der Haut im Verlaufe des Pockenprozesses,*" in *Medizingeschichte in unserer Zeit* (Ed. H. H. Eulner et al.), Stuttgart, 1971, 245–258.

"*La dottrina della circolazione del sangue e la Scuola Galileiana* 1636–61," in *Gesnerus,* 28 (1971), 7–34.

"*G. B. Morgagni und die Bedeutung seines 'De sedibus et causis morborum per anatomen indagatis',*" in *Gerard van Swieten und seine Zeit,* (Ed. E. Lesky u. A. Wandruszka), Wien-Köln-Graz, 1973, 128–136.

"*L'influence exercée sur la Médecine Clinique par les Sciences de Base développées par l'Ecole Galiléienne (Génération spontanée et 'Contagium Vivum' de la Gale),*" in *Clio Medica,* 8 (1973), 143–149.

Malpighi, Descartes, and the Epistemological Problems of Iatromechanism

FRANÇOIS DUCHESNEAU

[1] Luigi Belloni, *Opere scelte di Marcello Malpighi.* Turin: UTET, 1967; Howard B. Adelman, *Marcello Malpighi and the Evolution of Embryology,* 5 vols. Ithaca, N.Y. Cornell University Press, 1966.

[2] *Opere scelte di Malpighi,* pp. 512–513.

[3] Marcello Malpighi, *Anatome Plantarum.* London 1675, p. 1.

[4] *Opere scelte di Malpighi,* p. 510.

[5] Descartes, *Principes de la philosophie* in *Oeuvres* (eds. C. Adam and P. Tannery) Paris, 1973 (reprint) vol. IX–2, p. 163.

[6] Descartes, *Principles of Philosophy* in E. J. Haldane and G.R.T. Ross (eds. and trans.), *The philosophical works of Descartes,* 2 vols., Cambridge: University Press, 1967, vol. I, p. 301. Latin text in *Oeuvres,* vol. VIII–1, p. 328.

[7] Descartes, *Principes* in *Oeuvres,* vol. IX–2, pp. 324–325. Latin text vol. VIII–1, pp. 328–329.

[8] Descartes, *Oeuvres,* vol. XI, pp. 120–121.

[9] G. Canguilhem, *La formation du concept de réflexe aux XVIIe et XVIIIe siècles.* Paris: Presses universitaires de France, 1955, n. 30.

[10] One of Descartes's assertions in his letter to Buitendijck (1643) expresses the basic axiom of his physiological theory: "By the way, I do not admit various kinds of motion, but only local motion from place to place, which is common to all bodies, animate and inanimate alike" (A. Kenny, *Descartes Philosophical Letters,* Oxford: Clarendon Press, 1970, p. 146. Latin Text, *Oeuvres,* vol. IV, p. 65.

[11] Descartes, *Oeuvres,* vol. XI, p. 224.

[12] *Ibid,* pp. 224–225.

[13] *Ibid,* p. 226.

[14] Kenny, *Descartes' Philosophical Letters,* p. 243; *Oeuvres,* vol. V, p. 276.

[15] Kenny, p. 244; *Oeuvres,* vol. V, p. 277.

[16] Kenny, p. 244, *Oeuvres,* vol. V, p. 277.

[17] Descartes, *Principles in Haldane and Ross,* vol. I, pp. 115–116; *Oeuvres,* vol. VI, pp. 55–56.

[18] Descartes, *Principles* in Haldane and Ross, vol. I, p. 117; *Oeuvres,* vol. VI, p. 59.

[19] M. D. Grmek, "A Survey of the Mechanical Interpretations of Life", in *Biology,*

History, and Natural Philosophy, ed. by Allen D. Breck and Wolfgang Yourgrau, New York, Plenum Press, 1972, p. 187.

[20] On the problem of Descartes' genetics and his embryological theories, *cf.* Jacques Roger, *Les sciences de la vie dans la pensée française du XVIIIe siècle,* 2e éd., Paris, Armand Colin, 1971, pp. 140–154.

[21] Descartes, *Oeuvres* vol. IX–2, p. 124. Latin text, vol. VIII–1, p. 100.

[22] Descartes, *Discourse* in Haldane and Ross, vol. 1, pag. 109; *Oeuvres,* vol. VI, pp. 45–46.

[23] Descartes, *Oeuvres,* vol. XI, p. 277.

[24] *Cf.* M. D. Grmek, *"La notion de fibre vivante chez les médecins de l'école iatrophysique", Clio Medica,* V (1970), pp. 297–318.

[25] G. Canguilhem, *"Machine et organisme",* in: *La Connaissance de la Vie,* 2nd edition, Paris, Vrin, 1967, p. 112.

[26] *Cf.* the statement of Niels Stensen shortly after the *Traité de l'Homme* (1664) was published:

> *Pour ce qui est de monsieur Descartes, il connaissait trop bien les défauts de l'histoire que nous avons de l'homme, pour entreprendre d'en expliquer la véritable composition. Aussi n'entreprend-il pas de le faire dans son* Traité de l'Homme, *mais de nous expliquer une machine que fasse toutes les actions dont les hommes sont capables. Quelques-uns de ses amis s'expliquent ici autrement que lui; on voit pourtant au commencement de son ouvrage qu'il l'entendait de la sorte; et dans ce sens on peut dire avec raison, que monsieur Descartes a surpassé les autres philosophes dans ce traité dont je viens de parler. Personne que lui n'à expliqué mécaniquement toutes les actions de l'homme, et principalement celles du cerveau; les autres nous décrivent l'homme même; monsieur Descartes ne nous parle que d'une machine qui pourtant nous fait voir l'insuffisance de ce que les autres nous enseignent, et nous apprend une méthode de chercher les usages des autres parties du corps humain, avec la même évidence qu'il nous démontre les parties de la machine de son homme, ce que personne n'a fait avant lui. . .*

(from the *Discours sur l'anatomie du cerveau* (1665), in *Opera Philosophicà,* Vilhelm Maar, 1910, v. 2, pp. 7–8).

[27] François Duchesneau, *L'empirisme de Locke.* The Hague: Martinus Nijhoff, 1973.

Galileo's New Science of Motion

STILLMAN DRAKE

[1] A. Valdarnini, *Il metodo sperimentale da Aristotele a Galileo* (Asti, 1909), p. 63.

[2] Galileo, *Opere* I, p. 301; tr. I. E. Drabkin (Madison, 1960), p. 70. (*Opere* refers to the *Edizione Nazionale* of Galileo's works, ed. Antonio Favaro).

[3] *Opere* X, pp. 97–100.

[4] *Cf.* S. Drake, "Mathematics and Discovery in Galileo's Physics", *Historia Mathematica* I (1974), pp. 135–137. (Cited below as *Historia*).

[5] *Cf.* S. Drake, "Galileo's Discovery of the Law of Free Fall", *Scientific American,* 228 (1973), n.5, p. 89.

[6] *Historia,* pp. 139–143.

[7] *Historia,* pp. 143–148.

[8] *Opere* VIII, pp. 371–423.

[9] *Opere* I, pp. 299–300; II, p. 179; X, p. 170.

[10] Pseudo-Aristotle, *Questions of Mechanics,* n.1 (Loeb edition, pp. 337–339).

[11] Cf. S. Drake, "Galileo's Experimental Confirmation of Horizontal Inertia", *Isis* 64 (1973), pp. 291–299.

[12] *Opere* VIII, pp. 383–384; *cf.* also p. 243. A similar approach was used in Galileo's *Dialogue, Opere* VII, pp. 255–256; tr. S. Drake (Berkeley, 53), pp. 228–229.

[13] *Opere* VIII, pp. 281–282; tr. S. Drake (Madison, 1974) pp. 230–231.

[14] W. Wallace, "The Enigma of Domingo de Soto", *Isis* 59 (1968) pp. 384–401.

[15] Cf. M. Clagett, *The Science of Mechanics in the Middle Ages* (Madison, 1959) pp. 560–561.

[16] M. Varro, *De motu tractatus* (Geneva, 1584).

[17] G. B. Baliani, *De motu gravium solidorum et liquidorum* (Genoa, 1646) pp. 110–111.

[18] P. Mousnier (ed.), *Tractatus physica de motu locali . . . ex praelectionibus Honorato Fabry* (Lyons, 1646) pp. 88–90; cited below as Fabri. For the opinion of Descartes, see *Opere* XVII, p. 390 and XX, p. 612.

[19] Fabri, pp. 98 ff.

[20] *Opere* XVIII, pp. 11–13.

[21] H. Hertz, *Principles of Mechanics,* tr. Jones (Dover, 1956), p. 3.

Sources of Galileo's Early Natural Philosophy: Bibliographical Note

A. C. CROMBIE

The subject of this paper (which has been checked by Adriano Carugo and is presented as a result of our joint researches) is discussed in detail in our forthcoming book to be published as: A. C. Crombie, with the collaboration of Adriano Carugo, *Galileo and Mersenne: Science, Nature and the Senses in the Sixteenth and Early*

Seventeenth Centuries, 2 vols. This work is a considerably expanded version of our unpublished volume, *Galileo's Natural Philosophy* (1968), which was awarded the Galileo Prize and is deposited in the Domus Galilaeana, Pisa.

All citations of Galileo's published writings refer to *Le Opere di Galileo Galilei,* A. Favaro, ed., 20 vols. (Florence, 1890–1909): cited in the text as *Opere.* References are made to the major Latin edition Aristotelis Stagiritae *Omnia quae extant opera* . . . Averrois Cordubensis *In ea opera omnes qui ad nos pervenere commentarii* . . ., 11 vols. (*Venetiis apud Iuntas,* 1550–52); Galileo seems to have used a reprint of 1573–76.

Relevant secondary publications are C. M. Briquet, *Les filigranes: Dictionnaire historique des marques du papier dès leur apparition vers 1282 jusqu'en 1600,* a facsimile of the 1907 edition with supplementary material, ed. A. Stevenson, 4 vols. (Amsterdam, 1968); A. C. Crombie, "The Primary properties and secondary qualities in Galileo Galilei's natural philosophy", *Saggi su Galileo Galilei* (pre-print, Firenze, 1969: this series is in course of publication); Galileo Galilei, *Discorsi e dimostrazione matematiche intorno a due nuove scienze,* a cura di A. Carugo e L. Geymonat (Torino, 1958); E. Garin, *Scienza e vita civile nel Rinascimento italiano* (Bari, 1965); A. Procissi, *La collezione Galileiana della Biblioteca Nazionale di Firenze,* i, "Anteriori", "Galileo", compilata da Angiolo Procissi (Roma, 1959); William R. Shea, *Galileo's Intellectual Revolution* (London, 1972); William A. Wallace, "Galileo and the Thomists", in *St. Thomas Aquinas 1274–1974 Commemorative Studies* (Pontifical Institute of Mediaeval Studies, Toronto, 1974) 293–330: an innacurate note on p. 330 about the discovery of Galileo's early sources is to be corrected.

The study of the paper used by Galileo for these early autograph writings was begun by Adriano Carugo and extended with certain precisions by myself. All the paper is made with parallel wire lines 28-30 mm. apart, at right angles to which are fainter parallel textural lines about 1 mm. apart. The watermarks, always consistently related to the wire lines, appear on the folios at fairly regular intervals according to the foldings. By this criterion the writings may be grouped as follows:

1. On paper without watermark:
   Disputationes de praecognitionibus et de demonstratione (Biblioteca Nazionale Centrale di Firenze, *MSS Galileiani 27,* ff. 3-31; Procissi p. 106; Galileo, *Opere,* ix, 279-282, 291-292); Plutarch, *Opere morali (MSS Gal. 27,* ff. 34-42; Procissi p. 106; *Opere,* ix, 285-290); Sonetti (*MSS Gal. 27, f. 45; Procissi p. 107; G. O.* ix, 289-290); *La bilancetta* and *Tavola delle proporzioni delle gravità in specie dei metalli e delle gioie pesate in aria ed in aqqua (MSS Gal. 45,* ff.55, 60-62; Procissi p. 120; *Opere,* i, 215-20, 225-8); Fragment of Greek-Latin vocabulary *(MSS Gal. 70,* f.4; Procissi p. 148); *Dialogus de motu (MSS Gal. 71,* ff. 4-35; Procissi p. 151; *Opere,* i, 367-408); *Tractatus de motu (MSS Gal.* 71, ff.43-60; Procissi p. 151; *Opere,* i, 344-366).
2. On paper showing a mark CT or CL (*cf.* Briquet no.9553): *Tractationes de mundo et de caelo (MSS Gal. 46,* ff. 1-54; Procissi p. 123; *Opere,* i, 14-111).
3. On paper with watermark showing a backward-looking lamb with flag enclosed in a circle: Fig. 3 (Briquet no. 48): *Due lezioni all'Accademia fiorentina circa la figura, sito e grandezza dell'inferno di Dante* (1588; Bibl. Naz. Cent. di Firenze, *MSS Filza Rinuccini 21,* insertion 19, ff. 1-29; *Opere,* ix,

31-57); *Tractatūs de alteratione et de elementis (MSS Gal. 46,* ff. 57-100; Procissi p.123; *Opere,* i, 111-177, cf. 133); *Tractatus de motu (MSS Gal. 71,* ff. 115-124; Procissi p. 151; *Opere,* i, 326-340); *Isocratis ad demonicum admonitio (MSS Gal. 71,* ff. 125-132; Procissi p. 151; *Opere,* ix, 283-284).

4. On paper with watermark showing a forward-looking lamb with flag enclosed in a circle with a cross above: *Tractatus de motu (MSS Gal. 71,* ff. 61-104, 133-134; Procissi p. 151; *Opere,* i, 251-312, 341-343).

5. On paper with watermark showing a swan on three semicircles (Briquet no. 12550): *Tractatus de motu (MSS Gal. 71,* ff.105-114; Procissi p. 151; *Opere,* i, 312-326). The paper is whiter than that of the preceding and succeeding folios. There are linking marks H on ff. 104[v] and 105[r], and 7 on ff. 114[v] and 115[r]. Corrections and some repeated words throughout the *Tractatus de motu* suggest that Galileo was making a fair copy on different kinds of paper. In fact all the longer of these autograph writings show such mistakes.

6. On paper with watermark showing a ladder in a shield: *Dialogus de motu (MSS Gal. 46,* ff. 102-104; Procissi p. 123; *Opere,* i, 375-378, *cf.* 248); *Memoranda de motu (MSS Gal. 46,* ff. 102, 104-110; Procissi p. 123; *Opere,* i, 409-417); Italian-Latin vocabulary (*MSS Gal. 46,* f.112; Procissi p.123; *Opere* i, 246). *MSS Gal. 46,* f.113 continuing the vocabulary has a watermark showing a star above the shield with the ladder (Briquet no. 5926), and this appears also on blank ff.121-127.

The Role of Alchemy in Newton's Career

RICHARD WESTFALL

[1] David Brewster, *Memoirs of the Life, Writings, and Discoveries of Sir Isaac Newton,* 2nd ed., 2 vols. (Edinburgh, 1860), *2,* 300–302.

[2] The recent literature on Newton and alchemy, excluding articles that concern themselves primarily with his theory of matter, include R. J. Forbes, "Was Newton an Alchemist?" *Chymia* 2 (1949), 27–36; F. Sherwood Taylor, "An Alchemical Work of Sir Isaac Newton," *Ambix* 5 (1956), 59–84; D. Geoghegan, "Some Indications of Newton's Attitude towards Alchemy," *Ambix* 6 (1957), 102–106; Marie Boas and A. R. Hall, "Newton's Chemical Experiments," *Archives internationales d'histoire des sciences* 11 (1958), 113–152; Mary S. Churchill, *"The Seven Chapters,* with Explanatory Notes," *Chymia* 12 (1967), 29–57; P. M. Rattansi, "Newton's Alchemical Studies," in *Science, Medicine and Society in the Renaissance. Essays to Honor Walter Pagel,* ed. Allen G. Debus, 2 vols. (New York, 1972), *2,* 167–182. Most recent of all is the study certain to be recognized as authoritative by Betty Jo Dobbs, *The Foundations of Newton's Alchemy, or "The Hunting of the Greene Lyon";* Mrs. Dobbs' work is scheduled for publication by the Cambridge University Press in

1975. Although this remarkable work will be cited in more than one note below, I profited from it far more than the notes can indicate.

[3] Cambridge University Library, *Add. MS. 3975*. "Of Colours", pp. 1–22 (although "Of Colours" does contain a few notes from Boyle, it is an exception to the general characterization of these sections; it contains the first consistent exposition of Newton's theory of colors, the heart of his contribution to optics); "Of Cold, & Heate," pp. 25–41; "Rarity, Density, Elasticity, Compression &c," pp. 45–46; "Of fire, flame, ye heate & ebullition of ye heart & Divers mixed liquors. & Respiration," pp. 49–51.

[4] "Of Formes & Transmutations wrought in them," pp. 61–66; "Of Salts, & Sulpureous bodys, & Mercury. & Mettalls," pp. 71–100 (pp. 80–84 contain the results of Newton's own early experiments apparently in two groups—pp. 80–83 and pp. 83–84).

[5] "The medicall virtues of Saline & other Praeparations," pp. 159–174, continued on pp. 207–223 (broken down into a number of sub-headings such as "Of Spirit of wine," Volatile Salt," Alcalizate salts . . ."); "Of other Animall & Vegetable Substances," p. 177–182; "Medical observations," pp. 187–193.

[6] *Ibid.*, pp. 162, 209–223. Since the notes on Starkey are preceded by notes from Boyle's *Essays . . . of Effluviums*, published in 1673, the notes cannot be earlier than that year.

[7] "Gross Ingredients," First preparation," "3 Principles," "4 Elements," "Mercuries," "Sulphurs," "Salts," "Fires," pp. 227–241 (with no entries under any of these headings); "Of ye work wth common ⊙," pp. 243–244, continued on p. 261 (the notes from Philalethes work, published in 1678, and de Monte-Snyders are under this heading): "Of ye work wth artificial ⊙," "Times," "Proportions," "Hieroglyphicks," "Progress of ye Decoction," "Use of ye stone," "Miscellanies," pp. 245–257 (with no entries under any of these headings).

[8] Bodleian Library, *MS. Don. b. 15.*

[9] Notes on Basil Valentine, *Keynes MS. 64;* notes on Sendivogius, *Keynes MS. 19;* notes on Philalethes, *Keynes MSS. 51* and *52;* notes on Maier, *Keynes MS. 29. (Keynes MS. 29* contains the material that Newton used in his letter of 18 May 1669 to Francis Aston *(The Correspondence of Isaac Newton,* 4 vols. continuing, eds. H. W. Turnbull, J. F. Scott, and A. R. Hall [Cambridge, 1959 continuing] *1,* 11.) The *Keynes MSS.* are in King's College, Cambridge and are cited by their courtesy. All of the MSS. cited here are written in the hand of the late 60's. *MS. Var. 259,* in the Jewish National Library, Jerusalem, also contains, along with some other alchemical papers, early notes on Artephius, Flammel, Sendivogius, d'Espagnet, Augurello, Philalethes, Hermes *(Tabula Smaragdina),* and several pieces in the *Theatrum chemicum.*

[10] From a notebook in the Fitzwilliam Museum, Cambridge, n.p.

[11] *Ibid.*

[12] Early experiments: *Add. MS. 3975,* pp. 80–83; later experiments, pp. 83–84; still later, pp. 101–158, 267–283.

[13] *Keynes MS. 67.* Newton's corrections are found in the following items: "An unknown author, upon the philosophers stone," ff. 23v–26. (Ashmole attributes this poem to Pearce the Black Monke), and "The Vision of Sr George Ripley," f. 28. He numbered recipes on ff. 70–76v from 1 to 75, recipes on ff. 91–99^v^ from 1 to 106, and early paragraphs of "The briefe of Sr Edward Veres booke. August 18. 1610," (ff. 106v–108v) from 1 to 10. On f. 68v, which had been left blank in the collection, he wrote a paragraph with the title, "A liquor wherew^th^ a picture drawn upon marble doth sinke into it so y^t^ ye picture still appeare although ye stone bee broke or grownd away." *Keynes MS. 62* consists of Newton's copies of and notes on various pieces in *Keynes MS. 67:* "The work of an old Priest. viz: B" (*Keynes MS. 67,* ff. 19–19v), *"Ex Johanno Paupere" (Ibid.,* ff. 49–51), *"Ex Johanne Garlandio de p^r^paratione Elixaris" (Ibid.,* ff. 58v–59v). "The Philosophers water of might. p^r^ B" *(Ibid.,* 19v), "The briefe of Sr Edward Vere his book." (*Ibid.,* ff. 92-3, 106v–108v, two versions of the same treatise, which I have not bothered to collate), and a considerable number of recipes, two of which I located among the numbered recipes in ff. 70–77v before I decided that further identifications would be pointless.

[14] *Manna; Keynes MS. 33. De Scriptoribus Chemicis;* Stanford University, *MS. 6.*

[15] *Keynes MS. 51* contains notes on Philalethes' *Ripley Reviv'd;* not only is the hand early, but Newton's page references do not correspond to the pages in the published version of 1678. *Keynes MS. 52* is a copy of *An Exposition upon Sir George Ripley's Epistle to King Edward IV,* a part of the ultimate *Ripley Reviv'd.* Newton's copy contains sentences that are not in the published version. It has been collated with a MS. version, *Sloan MS. 633,* in the British Museum.

[16] *Manna,* in the hand of the mid 70's and with a note that Mr. F. gave it to Newton in 1675; *Keynes MS. 33. The Epitome of the treasure of health written by Eduardus Generosus Anglicus innominatus who lived Anno Domini 1562,* in the hand of the late 70's; *Keynes MS. 22.* John de Monte-Snyders, *The Metamorphosis of the Planets,* in the hand of the late 70's; Yale Medical Library. *Jodici a Rehe Opera Chymica,* in the hand of the late 70's and with copies of letters by John Twysden and A. C. Faber in 1673–1674; *Keynes MS. 50. Anno 1656. Serenissimi Principis Frederici Ducis Holsatiae et Sleswici &c communcatione sequens epistola me sibi vendicat, inaudita memorans. Veni et vidi,* in the hand of the late 70's or early 80's; *Keynes MS. 24.* William Yworth, *Processus mysterii magni philosophicus or An open Entrance to ye great Mysteries of ye Ancient Philosophies,* incomplete, in the hand of the 90's; *Keynes MS. 65.*

[17] Meheux to Newton, 2 March 1683; *Correspondence,* 2, 386. *Ibid.,* 4, 207–8, original in *Keynes MS. 26.* Newton wrote at least two drafts of this memorandum; they were separated at the auction, and the other draft is now in a private collection.

[18] In dating the alchemical papers, I have relied primarily on the evidence of Newton's hand, studied with all the care I could muster and compared with dated samples of his writing. To check my judgment I have used all of the internal evidence that the papers present, such as the dates in *Keynes MSS. 33* and *50,* notes of Mint business on *Keynes MSS. 13* and *56,* and especially the date of publication of books cited. Since he did have access to unpublished manuscripts, page references that can be collated with published versions have to be attended to. The work of Mundanus (at least as far as Newton was concerned), *De quintessentia*

philosophorum, published in 1686, offers major assistance. Newton cited it voluminously, and notes to Mundanus place a considerable number of the papers in the late, post-*Principia* period. I used the hand in which those papers were written to place others written in an indistinguishable hand in the same period.

In order to introduce some rough quantitative scale, I counted as one unit a single page of standard size written in a small hand. For the great bulk of the alchemical papers, Newton used sheets folded twice to make eight pages, each about nine inches by seven. Four such pages chosen at random in *Keynes MSS. 51* and *52* averaged 750 words each. They were written in the small hand of the late 60's. This was my basic unit. Because of the manifest crudity of the measure, I saw no point in counting words forever, and I proceeded from that basis by qualitative estimates. For a medium sized hand, I reduced the number of units by one-third; for a large hand by one-half. When he used folio sized pages, I multiplied by two. The limitations of these measurements should be obvious enough. Among other things, they treat each page of writing, whether it records Newton's own experiments, his comparisons of alchemical writers, or merely his copy of a manuscript, as equal. I trust that I am as aware of the limitations as the reader will be. Having come this far, however, I went on to compute a percentage of the whole for each of three periods. I do not insist on the exact figures in any way whatever. I do insist that the gross figures show that Newton devoted a great deal of time and energy to alchemy soon after the publication of the *Principia.*

By using the indications of length in the Sotheby catalogue, I estimate that I have been able to study over 80% of the total units on alchemy that Newton wrote. They include the chemical notes in his notebooks, his experimental notes, and the alchemical papers *per se.* I have not included the closely associated drafts of the *Queries* that are found in *Add. MS. 3970,* the papers connected with the *Opticks,* most of which derive from a later period. By my count, there are 1298 units among the papers I have studied and an estimated 255 units that I have not been able to see and date—a total of 1553 units. Hence the conclusion that in all he wrote nearly 1,200,000 words connected with alchemy, more than 1,200,000 words if the drafts of associated *Queries* are included. Of the 1298 units, 14% came from the late 60's, 33½% from the period 1674–1687, and 52½% from the early 90's. I have deliberately made these periods non-continuous, not to suggest breaks in Newton's study of alchemy, but to indicate my conviction that dating by these methods cannot be precise. In my eyes the large quantity from the early 90's is the most important. I have already indicated that the date of publication of Mundanus confirms the dating of a very large part of them. Newton might have written a few of those papers between the time of Mundanus' publication and the final completion of the *Principia.* There was simply not enough time for him to have written many of them in that short period when he was at work on the *Principia.* In dating them to the early 90's, I mean partly to indicate the partial interruption that the revolution and membership in the Convention Parliament involved, as suggested by his dated experimental notes, and partly to stress that this group of papers fell after the *Principia.*

[19] Consider sheer quantity, as with the alchemical papers. Volume I of Whiteside's edition covers the year 1664–1665 plus two short papers composed, after an intermission, in May, 1666, and a longer one composed, after another intermission,

in October 1666. The next four volumes cover the succeeding twenty years, and since most of the papers were composed in Latin, their publication in both the original and in translation roughly doubles their length. In crude terms, the twenty years following 1666 did not produce much more than twice the quantity of mathematical papers from the single year 1664-5. Moreover, a considerable proportion of them was devoted to elaborating ideas that dated from 1664–1665.

[20] *Add. MS. 3975,* p. 127: "Munday. June 26 1682." In 1682, 26 June did fall on Monday. p. 133: "Tuesday July 19." Neither in 1682 nor in 1683 did July 19 fall on Tuesday; since there is a date from 1683/4 on p. 135, those are the only two possible years. p. 149: "Friday May 23." Previous dated experiments place this in 1684 at the earliest, and later ones in 1685 at the latest; in 1684, 23 May did fall on Friday. p. 150: "Apr 26. 1686 Wednesday." In 1686, 26 April fell on Monday.

[21] *Ibid.,* inside back cover.

[22] Yale Medical Library. *Cf.* fn. 16.

[23] *Keynes MS. 23.* Corrections of the translation have the appearance of Newton's corrections of all his work.

[24] Jewish National Library, *MS. Var. 260.*

[25] The earlier notes are item 4 in Jewish National Library, *MS. Var. 259.* The later notes are in *Keynes MS. 19*; the recipe copied from the earlier notes is on ff. 1-1v.

[26] Babson College (formerly Babson Institute, Babson Park, Massachusetts), *MS. 417,* pp. 6–17. Courtesy of Babson College.

[27] *Cf. The Language of the Prophets,* in *Sir Isaac Newton: Theological Manuscripts,* ed. H. McLachlan, (Liverpool, 1950), pp. 119–126. I owe this suggestion to Mrs. Dobbs' work.

[28] For example, Keynes MS. 56, *Sententiae luciferae et Conclusiones notabiles,* contains six separate lists of equivalent names (ff. 7, 12, 12v).

[29] *Keynes MS. 46,* f. 1. I am convinced that I could collect several hundred similar passages from the papers. To support that assertion, I cite here a few more that I copied down before the repetivity of the task exhausted my patience: *Ibid.,* f. 1v; *Keynes MS. 19,* f. 3v (two examples); *Keynes MS. 20,* f. 5; *Keynes MS. 38,* ff. 4-4v, 9v; *Keynes MS. 40,* f. 23v; *The Regimen,* Burndy Library, n.p.

[30] *Babson MS. 420,* pp. 8–9, 12. *Cf. Ibid.,* pp. 11^a-13^a. *Keynes MS. 21,* ff. 16-16v, contains a similar paragraph of great interest because it expounds the meaning of the *Tabula Smaragdina,* one of the basic alchemical texts of all time:

> "The work is also thus described by Hermes. *Quod est superius est sicut id quod est inferius & quod est inferius est sicut id quod est superius* [*utrobique stellae et coelum*] *ad perpetranda miracula unius rei* [*materiae nostrae.*] *Pater ejus est Sol mater ejus Luna, portavit illud ventus* [*mercurialis*] *in ventre suo* [*quippe qui medium est conjungendi tincturas Solis et Lunae. Deinde infantis per hanc conjunctionem nati*] *Nutrix est terra* [*nostra dum per undecim gradus solvitur et a foecibus purgatur. Postea ad perfectionem deducitur &*] *Virtus ejus fit integra si* [*per decoctionem*] *versa fuerit in terram* [*fixam. Interea*] *Separabis terram* [*nigram fixam*] *ab igne* [*rubro fluido & volatili*] *subtile a spisso* [*per putrefactionem &*

> *destillationem] suaviter magno cum ingenio. Ascendit [post imbibitiones] a terra in coelum [per sublimationem ♀ ij exuberati] iterumque [per digestionem ad fixationem usque] descendit in terram, & recipit vim superiorum & inferiorum [spirituum & corporum vel Animalium vegitabilium & mineralum] Sic habebis gloria totius mundi. Haec est totius fortitudinis fortitudo fortis. Quia vincet omnem rem subtilem omnemque solidam petrabit* [*sic*–my bracket]."

[31] *Keynes MS. 19,* f. 3.

[32] Jewish National Library, *MS. Var. 259,* item 9.

[33] *Keynes MS. 54,* f. 2.

[34] *Keynes MS. 38,* f. 10v.

[35] *Keynes MS. 21,* f. 15v. *Cf.* an analogous passages in *Keynes MS. 54,* f. 4bis. At the end of this passage Newton cited Denis Zachaire, but in a manner that seems to use him to illustrate the assertion Newton has made rather than to rely on him as an authority for it.

[36] *Keynes MS. 30,* f. 1. My translation of Newton's Latin. The typed introduction to *Keynes MS. 35,* in my microfilm copy of the Keynes MSS., ascribes the identical list to the wrapper of *Keynes MS. 35,* but such a list is not there. I assume that the two lists are one and the same, somehow mistakenly transferred from *Keynes MS. 35* to *Keynes MS. 30.*

[37] *Keynes MS. 49,* f. 1.

[38] All but Index 3a are in *Keynes MS. 30.* Index 3a is in the Yale Medical Library. Index 3 is foliated. Indices 1a, 1, and 2, and the "*Supplementum Indicis Chemici,*" the drafts, are found after it, unfoliated.

[39] *Ibid.,* f. 58.

[40] *Ibid.,* f. 87.

[41] *Keynes MS. 32.*

[42] *Keynes MS. 21.* Beyond the close similarity of the content to the *Praxis,* the connection of the two is argued by the appearance of the Didier anagram ("*Dives sicut ardens S. C. Sanctus Didierus*") on the margin of the final page of the *Praxis.* (*Babson MS. 420,* p. 18^{a}.)

[43] *Keynes MS. 53.*

[44] *Babson MS. 420.* pp. 3–20 contain an entire treatise divided into five chapters. pp. 11^{a}–18^{a} contain a draft of chapters 4 and 5 which is not yet divided into two chapters.

[45] *Keynes MS. 48,* ff. 26–52, presents a coherent treatise named *Decoctio,* beginning with "Regimen Mercurij" and proceeding through the Regimens of Saturn, Jupiter, Luna, and (grouped together) Venus, Mars and the Sun. ff. 53–60 contains a preliminary draft of *Decoctio* without the title. The Burndy Library has two manuscripts entitled *The Regimen* and one *Of ye Regimen;* they resemble *Decoctio* in content as well as the use of "Regimen." *Of ye Regimen* was written over the remains of an incomplete receipt to John Day dated 11 September 1689; the hand of the

other two Burndy papers cannot be distinguished from its hand. In the *Index chemicus,* what appears to be a late entry under "Minerae" seems to be taken directly from a passage in *Decoctio,* and under "Mercurius duplex" Newton added some page numbers from *Decoctio* to a reference to Trevisan. This suggests that *Decoctio* was composed after the *Index* was nearly completed.

[46] Burndy Library, *The Regimen.* There is no way to distinguish the two papers with this title.

[47] *Keynes MSS. 40* and *41; Babson MS. 417.* I have compared a number of passages in the two Keynes MSS. with the *Index.* An entry from *Opus quintum* in *Keynes MS. 40* was copied directly into the section on Ferments in the *Index,* being entered on a blank verso opposite the original entries. There were a number of other passages in which the page citations in both MSS. corresponded exactly, including a passage on proportions in the *Opus octavum* that is identical to an entry under "Pondus" in the *Index. Keynes MS. 41* also has a passage on proportions that appears, from its details, to have been based on another entry under "Pondus" in the *Index.*

[48] *Extractio et rectificatio animae* (not always with that identical title) in order of composition as far as I can establish it: *Keynes MS. 41,* ff. 2–3; *Babson MS.* 417, Part II, p. 13; Part II, pp. 18–20; Part II, pp. 3–10 (two drafts), Part I, pp. 21–28. *Extractio et rectificatio spiritus* (not always with that identical title) in order of composition as far as I can establish it: *Keynes MS. 40,* f. 3; *Keynes MS. 41,* ff. 1–2; *Babson MS. 417,* Part II, pp. 17–18; Part I, pp. 18–20. *Elementorum qualitates: Babson MS. 417,* Part II. pp. 11–12; Part I, pp. 29–31. *Salis imbibitio & sublimatio in terram albam foliatam: Keynes MS. 40,* ff. 9–12v; *Keynes MS. 41,* ff. 3v–5. *Solutio sicca & humida metallorum . . .: Keynes MS. 40,* ff. 29–30v; *Keynes MS. 41,* ff. 11v–20. Item 66 in the Sotheby catalogue speaks of "Notes on the Operations, 1, 2, and 6–9 . . . with very many alternations and re-writings." I think that there are in fact more corresponding drafts among *Keynes MSS. 40* and *41* and *Babson MS. 417.* After a certain amount of comparison of MSS. of this sort, one has to stop to preserve his sanity.

[49] *Keynes MS. 40,* f. 19—one example of which many could be given.

[50] *Keynes MS. 41,* f. 3.

[51] Hermes: *Keynes MSS. 27* and *28.* Maria the Jewess: *Keynes MS. 45.* Lull (or Raymund, as Newton referred to him): *Keynes MS. 47.* Ripley: *Keynes MSS. 53* and *54.*

[52] *Keynes MS. 13.*

[53] *Keynes MS. 46.*

[54] *Sendivogius explained; Keynes MS. 55.* The content of this treatise reflects themes of Newton's other papers to the point that it is hard to believe it is not his composition. Nevertheless, the lack of revisions and the absence of drafts raise doubts. Newton was not one to write down a work in finished form on the first try. There are a few corrections in Sendivigus explained, but such as suggest copying errors; they are certainly not the sort of revisions familiar to every student of Newton's papers. *Ripley Expounded; Keynes MS. 54,* ff. 1–6.

[55] *Keynes MS. 66.*

[56] *Add. MS. 3975,* pp. 80–83. Further experiments on pp. 83–84 appear to be a distinct set somewhat later in time.

[57] *Keynes MS. 18.* Since it is only two folios in length, I shall not bother to cite the location of individual passages that I use. For a more detailed argument in support of Newton's authorship, see Mrs. Dobbs' work. The section of my paper on the *Clavis,* together with the preceding paragraph about Newton's notes on making reguli, is drawn directly from Mrs. Dobbs' work, and in citing the *Clavis,* I use her English translation (Appendix D). I have also profited from extensive correspondence with her. I am delighted that this outstanding piece of scholarship will soon be placed before the community of Newton Scholars.

[58] "Philosophical ☿ wch wil dissolve it selfe & congeale ye bodies, o^{r} Hermaphrodite is made out of ye Reg of ♂ & common ☿ fermented together by ye mediation of Diana's Doves." *Keynes MS. 34,* f. 1. *Cf. ibid.,* f. 3v; *Keynes MS. 35,* sheet 7; Burndy Library, *The Regimen; Babson MS. 420,* pp. 3–4, 9–10, 13–14.

[59] There are two sets of experimental notes, *Add. MS. 3973* and those in the notebook, *Add. MS. 3975.* The majority of the notes in *MS. 3973* were copied, sometimes with emendations, into notebook *3975,* although not every note in *3975* has a counterpart in *3973.* The first group of notes in *3975* (pp. 101–105) have no date. A new set, which begins on p. 106, copied virtually word for word notes in *3973* dated February 1679/80. Since there was a change of ink at the bottom of p. 104, the early set may in fact have been two sets, and that would bear upon the date of the earliest ones. Nevertheless, 1678 appears a reasonable conjecture.

[60] *Add. MS. 3973,* ff. 12, 36v. *Add. MS. 3975,* pp. 104, 106, 109.

[61] *Cf.* a note from Sendivogius: "*Scito enim metallorum vitam esse ignem dum adhuc in suis mineris sunt et mortem etiam ignem fusionis videlicet.*" *Keynes MS. 48,* f. 15. *Cf. Keynes MS. 55,* f. 17.

[62] *Keynes MS. 30,* ff. 25, 48v; *Keynes MS. 48,* f. 20.

[63] *Add. MS. 3973,* ff. 5; 25v. *Keynes MS. 38,* ff. 8, 9, 11. *Keynes MS. 41,* f. 3. *Babson MS. 417,* pp. 21–22. There are innumerable other examples in the papers.

[64] *Add. MS. 3975,* p. 83.

[65] *Babson MS. 420,* p. 11.

[66] *Keynes MS. 41,* ff. 16–16v.

[67] *Keynes MS. 35,* sheet 7. *Keynes MS. 40,* f. 31. *Keynes MS. 55,* ff. 2–3v. *Keynes MS. 56,* ff. 1v–2. Note also the reference in the *Praxis,* cited above, fn. 65.

[68] *Add. MS. 3975,* pp. 133–134.

[69] *Keynes MS. 35,* sheet 7. *Keynes MS. 40,* ff. 27v–28.

[70] *Add. MS. 3973,* f. 44.

[71] *Keynes MS. 30,* ff. 33, 61. *Keynes MS. 38,* f. 1. *Keynes MS. 53,* f. 2v. *Babson MS. 420,* p. 15.

[72] *Add. MS. 3973,* f. 19. *Add. MS. 3975,* pp. 151–152.

[73] "May 10 1681 *intellexi Luciferam ♀ et eandem filiam ♄* [ni], *& unam columb*[rum]. May 14 *intellexi* -∈. May 15 *intellexi Sunt enim quaedam ☿* [ij] *sublimationes &c ut & columbam alteram: nempe Sublimatum quod solum foeculentum est, a corporibus suis ascendit album, relinquitur foex nigre in fundo, quae per solutionem abluitur, rursusque sublimatur ☿* [ius] *a mundatis corporibus donec foex in fundo non amplius restet. Nonne hoc sublimatum depuratissimum sit*→*?" *Add. MS. 3975,* p. 121.

[74] "May 18 *Ideam solutionis perfeci Nempe aequalia duo salia elevant ♄ Dein hic elevat lapidem, nec non cum Jove malleabili conjunctus fit ⚹ idque in tali proportione ut ♃ sceptrum apprehendat. Tunc aquila ♃* [em] *attollet. Potest dein ♄ sine salibus in ratione desiderata conjugi ne ignis praedominetur. Denique ☿ sublim. & * praeparat feriunt cassidem, & menstruum omnia attollit.*" *Ibid.,* p. 122.

[75] *Add. MS. 3973,* f. 17.

[76] *Babson MS. 420,* p. 14[a]. *Keynes MS. 40,* f. 27. *Keynes MS. 41,* ff. 14f. *Cf. Keynes MS. 30,* ff. 48v, 49v, and *Keynes MS. 35,* sheet 6. For the image of Neptune and the Trident possibly referred to in the passage of 14 May, see also *Add. MS. 3973,* f. 12. For the image of the helmet of Mercury used on 18 May, see *Babson MS. 420,* p. 14[a].

[77] *Keynes MS. 41,* ff. 3v-4. Although I have only the one example ready at hand, I saw innumerable references to "*nostrum sal armoniacum*" in the papers.

[78] *Keynes MS. 59,* f. 1v.

[79] "Friday May 23 *Jovem super aquilam volare feci.*" *Add. MS. 3975,* p. 149.

[80] Eirenaeus Philalethes, "An Exposition upon Sir George Ripley's Preface," in *Ripley Reviv'd,* (London, 1678), p. 28. After Philalethes' figure caught my eye, I began to realize that it had caught Newton's about three hundred years earlier. I can record at least five citations of it, and I am sure that I saw it at least twice in the *Index chemicus* before I knew I was collecting instances. *Keynes MS. 34,* f. 1v. *Keynes MS. 35,* sheet 4. *Keynes MS. 48,* ff. 16–16v. *Keynes MS. 51,* f. 1v. *Babson MS. 420,* p. 8. Amateur psychoanalysts may wish to interpret Newton's attraction to this figure; I must refrain since I would be analyzing myself as well.

[81] *Add. MS. 3975,* p. 119. *Cf. Keynes MS. 35,* sheet 7.

[82] Newton to Boyle, 28 February 1678/9; *Correspondence, 2,* 293–294. *De aere,* in *Unpublished Scientific Papers of Isaac Newton,* eds. A. R. and M. B. Hall, (Cambridge, 1962), pp. 216–220.

[83] It should go without stating by this point that Mrs. Dobbs' work makes a considerable stride toward reversing this situation.

[84] *Babson MS. 420,* p. 18[a]. This passage comes from a draft of the *Praxis.* The final version (p. 17) toned the statement down somewhat; it did not abandon its general assertion, however.

[85] *Add. MS. 3973,* f. 29.

[86] *Cf. Babson MS. 417,* Part II, p. 14, where Newton substituted the Paracelsian terms, body, soul, and spirit, for Lull's Aristotelian terms, earth, water, and air.

87 *Add. MS. 3996,* ff. 88–135.

88 *Unpublished Papers,* pp. 140–148.

89 *Ibid.,* p. 148.

90 From the *Turba: "Et scitote quod secretum operis ex mare et femina constat, hoc est agente et patiente." (Keynes MS. 25,* n.p.) From Ferrar the Monk: *"Nam corpus immundum sine fermento quod est ejus anima, mortuum est et immobile. . . . Fermentum autem quod corpori intruditur est ejus anima." (Keynes MS. 38,* f. 11v.) From Trevisan: *"Opus nostrum fit ex una radice & duabus substantijs mercurialibus, assumptis omino crudis, tractis ex minera, puris et mundis, conjunctis per ignem amicitiae ut materia requirit, coctis assidue donec ex duobus fiat unum: in quo uno corpus spiritus et iste corpus facta sunt a commixtione." (Keynes MS. 48,* f. 18v.) From the *Great Rosary,* In the unversal way, the alchemist *"activa junguntur passivis* . . ." (*Ibid.,* f. 24v.) From Philalethes: The sophic mercury is an hermaphrodite "including in it both an active & passive principle . . ." By itself in the fire it coagulates into gold or silver. "So then y^{e} 🜍 wth wch this ☿ is impregnated is in truth volatile indigested ☉ & therefore it passeth into ☉ by bare digestion." (*Keynes MS. 40,* f. 22v.) In *Sendivogius explained,* if Newton was its author, he repeated the same themes to himself: ". . . things themselves or their essences ly hid under shadows clothed wth certain sensible Elementary coverings." (*Keynes MS. 55,* f. 2.) "Take away ye shadow or scoria from y^{e} * of ♂ & ♀ & wash away ye cloud, that is ye black 🜍 from y^{e} * of ♂ & ♀ & ye filth from ye philosoph. ☿ & you will see ye point of ye magnet answering to every center of ye rays of ☉ & of ye earth or * of ♂ & ♀ . . . . The shadow of ye seed of nature is ye blackness of ye scoria & feces of y^{e} * of ♂ & ♀ ." (*Ibid.,* f. 11v.)

91 Burndy Library. Since the MS. has no foliation, I shall not try further to specify my citations from it. I cite this MS. and the *Regimen MSS.* that I have referred to more briefly before by courtesy of the Burndy Library.

92 *Keynes MS. 12,* which appears to stem from much the same time, propounds similar views in a series of short "Prepositions" (f. 1–2.):

> *Omnes species sunt ex una radice . . .*
> *Idem et unicus est agens vitalis per omnia quae in mundo sunt diffusus*
> *Estque spiritus mercurialis subtilissimus et summè volatilis per omnia loca dispersus.*
> *Hujus agentis eadem est methodus generalis operandi in omnibus, nempe modico calore excitatur ad agendum, fugatur magno, et proposito substantiarum aggregato prima ejus actio est putrefacere & in chaos confundere, deinde ad generationem procedit . . . .*
> *In maxima copia sub forma metallica reperitur in Magnesia*
> *Et ex hac unica radice sunt omnes metallorum species . . ."*

93 Collins to James Gregory, 19 October 1675; *Correspondence, 1,* 356.

94 *Ibid., 1,* 364.

95 *Add. MS. 3973,* f. 42. *Add. MS. 3975,* p. 281. *Cf. Add. MS. 3973,* ff. 6, 7v, 36, and *Add. MS. 3975,* pp. 103, 156–157, 277.

96 *Ibid.,* pp. 104–105. He had found similar ideas in Sendivogius: *"Magnes est nos-*

ter quem in praecedentibus Chalybem esse dixi Plumbum autem dicunt magnetem quia ☿ *ius ejus attrahit semen Antimonij sicut magnes Chalybem." (Keynes MS. 19,* f. 1.) Also in de Monte-Snyders: "Salt is the magnet of the soul . . . And where sulphur is extracted, there also a most noble mercury is extracted, since one is the magnet of the other & they are not easily parted." *(Keynes MS. 41,* f. 13) *Cf. Keynes MS. 19,* f. 3; *Keynes MS. 35,* sheet 7; *Keynes MS. 48,* ff. 37, 43; *Babson MS. 417,* pp. 19, 22, 23; Burndy Library, *The Regimen.*

97 *Add. MS. 3973,* ff. 13, 21. *Add. MS. 3975,* pp. 108–109. *Cf. Add. MS. 3973,* ff. 10v, 22v, 32v, and *Add. MS. 3975,* p. 133.

98 *Ibid.,* p. 107. *Cf. Ibid.,* p. 120.

99 *Add. MS. 3973,* f. 12. *Add. MS. 3975,* p. 107. *Clavis; Keynes MS. 18,* ff. 1–1v. A passage from Maier spoke of sons rejoicing to wash themselves in the maternal spring, which also served as food and drink to them. *"Similia enim a similibus attrahuntur et in seinvicem, nempe nutrimentum in nutritum transeunt." (Keynes MS. 49,* f. 2.) *Cf. Keynes MS. 19,* ff. 3–3v; *Keynes MS. 28,* f. 6v; *Keynes MS. 38,* f. 1; *Keynes MS. 48,* f. 23.

100 He also found this concept in the alchemists he studied. *Cf. Keynes MS. 35,* sheet 10; *Keynes MS. 41,* f. 9v.

101 *Add. MS. 3973,* ff. 35–37.

102 *Opticks,* based on 4th ed., (New York, 1952), pp. 397–400. Recently J. E. McGuire, relying on a small number of separated passages in manuscripts, has advanced the extraordinary conceit that Newton's active principles were an independent level of being. ("Force, Active Principles, and Newton's Invisible Realm," *Ambix, 15* (1968), 154–208.) This sort of confusion arises from the mistaken notion that Newtonian science can be studied in his speculations alone.

103 *Principia,* ed. Florian Cajori, (Berkeley, 1960), p. 325.

104 *Conclusio; Unpublished Papers,* p. 328. A draft preface to the first edition of the *Principia; ibid.,* p. 303. A draft of proposed changes to Book III from the early 90's; *ibid.,* p. 315. A similar draft; *Add. MS. 3965.6* f. 266v.

105 *Conclusio; Unpublished Papers,* p. 333.

106 A MS. connected with the first edition of *Opticks; Add. MS. 3970.3,* f. 337.

107 *Isaac Newton's Papers & Letters on Natural Philosophy,* ed. I. Bernard Cohen, (Cambridge, Mass., 1958), p. 258.

108 A MS. connected with the Latin edition of *Opticks; Add. MS. 3970.3,* f. 292.

109 *Papers & Letters,* p. 257.

110 *Keynes MS. 30,* ff. 66–69.

111 *Add. MS. 3973,* ff. 13–13v.

112 *Add. MS. 3975,* p. 271.

113 *Ibid.,* p. 143.

114 *Add. MS. 3973,* f. 36v.

[115] *Add. MS. 3975*, p. 125.

[116] *Keynes MS. 58*, f. 7.

[117] *Add MS. 3973*, ff. 30v–31. *Cf. ibid.*, f. 38: "In sublimate of ♁ 6 parts of * caries up 3 parts of ♁ & by letting go a good quantity of sp^{t} of * loses 1/6 of its weight so that in ye sublimate of ♁ there is but 5 of * to 3 of ♁ . 6 pts of * gives 6½ of sublimate besides 1½ of yellow flowers. This sublimate 3^{pts} sublimed from ⊕ 2^{pts} carries up 1 of y^{e} ⊕ or 1 1/12 & by letting go much sp^{t} of * becomes a sublimate weighing 3¼ whereof 2/5 is ♁ & 3/5 is * + ☿ "*Cf. Ibid.*, ff. 32–32v, 39v–40, and *Add. MS. 3975*, p. 158.

[118] *Add. MS. 3973*, f. 38v.

[119] *Keynes MS. 40*, ff. 19v, 20. *Keynes MS. 41*, f. 15v. Burndy Library, *The Regimen. Babson MS. 417*, p. 35.

[120] *Keynes MS. 66*, f. 4v. I cannot be certain that this *MS.* is Newton's composition. On the bottom of f. 1 of *Keynes MS. 91*, which consists of alchemical fragments not in his own hand, Newton entered a sentence identical to the second sentence in the quotation from *Keynes MS. 66*. The *Praxis* contains a similar juxtaposition of arcane imagery with quantitative precision: "This rod & ye male & female serpents joyned in ye proportion of 3, 1, 2 compose ye three headed Cerberus wch keeps ye gates of Hell." (*Babson MS. 420*, p. 12.)

[121] *Conclusio; Unpublished Papers*, p. 333.

[122] *Keynes MS. 38*, f. 8.

[123] *Babson MS. 420*, pp. 14^{a} (a draft) and 13. From Newton's citation the letter can easily be identified as that of 4 May 1693. (*Correspondence, 3*, 265–266.)

[124] *Keynes MS. 13. Keynes MS. 56*. The note on *Keynes MS. 56* was jotted down on a blank space on the first sheet of a paper that was manifestly already written. It testifies only that Newton was looking at it after the move to London. Perhaps the same is true of *Keynes MS. 13*, although the case is less clear. The Mint business is found on f. 2v, the back of a sheet folded once, the first three sides of which have lists of alchemical authors. Since the note is not upside down, this side represented the back page to Newton when he wrote it, and he was not in the habit of using the back side of his familiar folded sheets when the front was available.

Newton, a Sceptical Alchemist?

PAOLO CASINI

[1] In Allen G. Debus (ed.), *Science, Medicine and Society in the Renaissance. Essays to Honor Walter Pagel*. New York: Science History Publications, 1972, Vol. II, pp. 183–198.

[2] R. S. Westfall, *Force in Newton's Physics*. London: Macdonald, 1971, p. 367.

[3] R. S. Westfall, "Newton and the Hermetic Tradition" (note 1), p. 186.

Hermeticism, Rationality and the Scientific Revolution

PAOLO ROSSI

[1] P. K. Feyerabend, "Problems of Empiricism, II", in R. G. Colodny (ed.), *Nature and Function of Scientific Theories,* Pittsburgh, 1970, pp. 275–353; P. K. Machamer, "Feyerabend and Galileo: the Interaction of Theories and the Reinterpretation of Experience", in *Studies Hist. Philos. of Science,* IV (1973) 1, pp. 1–46.

[2] J. E. McGuire, "Newton and the Demonic Furies: Some Current Problems and Approaches in History of Science", in *History of Science,* XI (1973), pp. 21–41; P. Rossi, *"Considerazioni sulla storia delle scienze",* in *Storia e Filosofia,* Turin: Einaudi, 1969, pp. 238–239.

[3] *The Works of Francis Bacon* (eds. R. L. Ellis, J. Spedding, D. D. Heath), London, 1887–1892, III, pp. 164–165.

[4] M. Boas Hall (ed.), *Nature and Nature's Laws. Documents of the Scientific Revolutions,* New York, 1970, p. 1.

[5] A. Koyré, *Études d'histoire de la pensée scientifique,* Paris, 1966, pp. 1–5.

[6] *The Works of F. Bacon,* IV, p. 294 (*De Augmentis,* Latin text: I, p. 496).

[7] P. Gassendi, *Syntagma,* in *Opera Omnia,* Lyons, 1658, I, 122B–123A.

[8] Gassendi, *Ad librum D. Edoardi Herberti Angli De Veritate Epistola,* in *Opera Omnia,* III, 413B–414A.

[9] *Oeuvres de Descartes* (eds. Ch. Adam et P. Tannery), Paris, 1897–1909, IX, p. 21 (*Principia*).

[10] M. Mersenne, *Harmonie Universelle,* Paris, 1636, Vol. III, p. 8 of the *"Nouvelles observations physiques et mathématiques".*

[11] Th. Hobbes, *The English Works* (ed. by W. Molesworth), London, 1839–1845, VI, pp. 183–184 (*Six Lessons to the Professors of the Mathematics*); *Opera Philosophica quae latine scripsit* (ed. W. Molesworth), London, 1839–1845, II, pp. 92–94 (*De Homine,* II, 10).

[12] G. B. Vico, *The New Science* (eds. G. Bergin and M. H. Fisch), New York, 1961, pp. 52–53. The texts of *De nostri temporis studiorum ratione* and *De antiquissima Italorum sapientia* are in *Opere* (ed. F. Nicolini), Milan, 1953, pp. 293, 307. For a more general discussion see: A. Mansion, *Introduction à la physique aristotélienne,* Paris, 1945, pp. 198–202; 228–234; 256–257; P. M. Schuhl, *Machinisme et philosophie,* Paris, 1947; A. Koyré, *"Les Philosophes et la machine",* in *Études d'histoire de la pensée philosophique,* Paris, 1961, pp. 279–309; P. Rossi, *Philosophy, Technology and the Arts in the Early Modern Era,* New York, 1970, pp. 136–145; W. Leiss, *The Domination of Nature,* New York, 1972, pp. 45–97; A. Child, *Making and Knowing in Hobbes, Vico and Dewey,* Los Angeles, 1953; R. Mondolfo, *Il verum-factum prima di Vico,* Naples, 1969; T. Gregory, *Scetticismo e empirismo: studio su Gassendi,* Bari, 1961.

[13] A. R. Hall, "The Scholar and the Craftsman in the Scientific Revolution", in M. Clagett (ed.), *The Critical Problems in the History of Science,* Madison, 1962, pp. 3–23.

[14] F. A. Yates, "Bacon's Magic", in *The New York Review of Books,* 29 February, 1968; *The Rosicrucian Enlightenment,* London, 1972, p. 119; P. M. Rattansi, "The Social Interpretation of Science in the Seventeenth Century", in P. Mathias (ed.), *Science and Society: 1600–1900,* Cambridge, 1972, pp. 13, 17.

[15] H. B. White, *Peace amongst the Willows: the Political Philosophy of Francis Bacon,* The Hague, 1968; Yates, *Rosicrucian Enlightenment,* pp. 118–129.

[16] *The Works of Fr. Bacon,* II, p. 13 *(Historia Naturalis et Experimentalis ad condendam Philosophiam).* J. R. Ravetz, "Francis Bacon and the Reform of Philosophy" in A. Debus (ed.), *Science, Medicine and Society in the Renaissance,* New York, 1972, II, p. 115, states that "Giordano Bruno receives no mention whatever in Bacon's reflective writings". In the *Index to the Philosophical Works* in the Spedding edition the name of Bruno is inexplicably omitted.

[17] See Copernicus, *De revolutionibus,* Nuremberg, 1543, pp. 9–10; W. Gilbert, *De Magnete,* trans. by P. Fleury Mottelay, New York, 1958, pp. 309–310; *Le opere di Galileo Galilei* (ed. A. Favaro), Florence, 1890–1909, V, p. 301; *Oeuvres de Descartes,* X, pp. 217–218; L. Couturat, *Opuscules et fragments inédits de Leibniz,* Paris, 1903, p. 511 *(Introductio ad Encyclopediam Arcanam).*

[18] M. B. Hesse, "Reason and Evaluation in the History of Science", in M. Teich and R. Young (eds.), *Changing Perspectives in the History of Science,* London, 1973, pp. 129–130, 137–138, 147; P. M. Rattansi, "Some Evaluations of Reason in Sixteenth and Seventeenth-Century Natural Philosophy", *ibid.,* pp. 148–166.

[19] Rattansi, *Some Evaluations,* pp. 148–149, 152, 155–156, 166.

[20] This basic ambiguity is never clearly confronted.

[21] T. S. Kuhn, "Reflections on my Critics", in I. Lakatos and A. Musgrave (eds.), *Criticism and the Growth of Knowledge,* Cambridge, 1970, p. 263; I. Lakatos, "Falsification and the Methodology of Scientific Research Programmes", *ibid.,* pp. 91–126; P. K. Feyerabend, "Consolation for the Specialist", *ibid.,* pp. 197–230.

[22] Lakatos, "Falsification and Methodology", p. 178.

[23] Feyerabend, "Consolation for the Specialist", pp. 216, 218, 228.

[24] Machamer, "Feyerabend and Galileo", p. 36.

[25] G. Gentile, *Teoria generale dello Spirito come Atto puro,* Firenze, 1958, pp. 219–224.

[26] F. Manuel, *A Portrait of Isaac Newton,* Cambridge (Mass.), 1968. For a stimulating discussion see McGuire, "Newton and the Demonic Furies".

[27] Lakatos, "Falsification and Methodology", p. 93.

[28] J. R. Ravetz, "Tragedy in the History of Science", in Teich and Young, *Changing Perspectives,* p. 122.

The Mathematical Revolution of the Seventeenth Century

RENÉ TATON

[1] R. Taton, *"Les origines de la géometrie projective", Notes du 2ᵉ Symposium International d'Histoire des Sciences* (Pisa-Vinci, 16–18 June 1958), Florence, 1960, pp. 248–255.

[2] D. T. Whiteside, "Pattern of Mathematical Thought in the Later Seventeenth Century", *Archive for History of Exact Sciences,* vol. I, n. 3, 1961, p. 180.

[3] A. P. Youschkevitch, *"Sur la Révolution en Mathématique des temps modernes", Acta Historiae Rerum Naturalium necnon Technicarum,* issue 4, Prague 1968, p.6.

Editors' Acknowledgement

The papers in this volume were first discussed at a Symposium organized by the *Gruppo Italiano di Storia della Scienza* under the chairmanship of Professor Vasco Ronchi at the Villa Malaparte on the island of Capri in April 1974. We wish to express our deep thanks to the members of the *Fondazione Giorgio Ronchi,* and especially to Professor and Mrs. Vasco Ronchi, Mrs. Maria Burchi Suckert, Professor Lucia Rositani Ronchi and Professor Laura Abbozzo Ronchi for their warm hospitality. We are also indebted to Miss Daniela Galassini of the *Istituto Nazionale di Ottica* of Florence for her assistance in organizing the Symposium and to Mrs. Marta Balduini Zangheri for her skill in deciphering and typing the various manuscripts.

We were privileged to receive the invaluable help of Mr. Neale Watson who attended the Symposium and advised us on matters concerning publication.

Finally, we are pleased to acknowledge the support of the Istituto e Museo di Storia della Scienza of Florence, the Canada Council and the Canadian Cultural Institute in Rome.

MARIA LUISA RIGHINI BONELLI
Istituto e Museo di Storia della Scienza
Florence, Italy

WILLIAM R. SHEA
McGill University
Montreal, Canada

Library of Congress Cataloging in Publication Data
Main entry under title:

Reason, experiment, and mysticism in the scientific
revolution.

Papers presented at a symposium held in Capri in
April 1974 and sponsored by the Gruppo italiano di
storia della scienza.
1. Science—History—Europe—Congresses.
I. Righini Bonelli, Maria Luisa, ed. II. Shea,
William R., ed. III. Gruppo italiano di storia delle
scienze
Q127.E8R4 509′.4 74–16495
ISBN 0–88202–018–8